Shipboard Application of Infr
Signature Suppression Technologies

AV Satyanarayana Murty

Shipboard Applications of Infrared Signature
Suppression Technologies

First Edition February 2024

Written by AV Satyanarayana Murty

CONTENTS

Page No.

LIST OF FIGURES

LIST OF TABLES

NOMENCLATURE

CHAPTER   1   INTRODUCTION   **1-18**

    1.1   BACKGROUND   1

    1.2   NAVAL SHIP SIGNATURES   2

        1.2.1   Acoustic Signature   2

        1.2.2   Magnetic Signature   3

        1.2.3   Wake Homing Signature   3

        1.2.4   Extremely Low Frequency Emissions (ELFE)   4

        1.2.5   Radar Cross Section (RCS)   4

        1.2.6   Laser Cross Section   4

        1.2.7   Air Borne Noise Signature   4

        1.2.8   Thermal / Infrared (IR) Signature   5

    1.3   Electromagnetic Radiation   5

    1.4   Infrared (IR) Radiation   6

        1.4.1   Sub Division of Infrared Wavebands   6

        1.4.2   Atmospheric Transmission of Infrared Radiation   7

        1.4.3   Applications of IR based Technologies   8

    1.5   SOURCE OF SHIP INFRARED SIGNATURE   8

        1.5.1   Exhaust Duct Assembly   9

        1.5.2   Exhaust Plume   10

        1.5.3   Hull and Assorted Surfaces   11

    1.6   IR SIGNATURE SUPPRESSION TECHNOLOGIES   12

        1.6.1   Cooling of Exhaust Plume and Duct Surfaces   12

		1.6.2	Cooling of Hull Surfaces	13
		1.6.3	Challenges in Design of IRSS for Ship Applications	13
	1.7	PROBLEM DEFINITION		14
	1.8	MOTIVATION		14
	1.9	OBJECTIVES		15
	1.10	ORGANISATION OF THE BOOK		16
CHAPTER	**2**	**LITERATURE REVIEW**		**19-40**
	2.1	INTRODUCTION		19
	2.2	SHIP IR SIGNATURE SUPPRESSION TECHNOLOGIES		19
	2.3	EJECTOR DESIGN FOR SHIP EXHAUST SYSTEMS		24
	2.4	ENTRAINING DIFFUSER DESIGN FOR SHIP EXHAUST SYSTEMS		31
	2.5	STRUCTURAL & FLUID PARAMETERS FOR IRSS		36
	2.6	TECHNOLOGY GAPS FROM LITERATURE		38
	2.7	SCOPE OF PRESENT RESEARCH		39
	2.8	CONCLUSION		39
CHAPTER	**3**	**MATERIALS AND TOOLS**		**41-63**
	3.1	INTRODUCTION		41
	3.2	MATHEMATICAL MODELLING		41
		3.2.1	Building Mathematical Models EDSA and ATPM	42
		3.2.2	Sensitivity Analysis	42
	3.3	NUMERICAL SIMULATION		43
	3.4	MODELLING		43
	3.5	COMPUTATIONAL FLUID DYNAMICS		44
		3.5.1	Creating Domain	46
		3.5.2	Meshing	46
		3.5.3	Meshing Models	47
		3.5.4	Volumetric Controls	50
		3.5.5	Reference Values of Meshing	50
		3.5.6	Meshing Process	51

		3.5.7	Defining Physics Models	51
		3.5.8	Defining Boundary Conditions	52
		3.5.9	Governing Equations	53
		3.5.10	Run the Simulation	53
		3.5.11	Post Processing	53
	3.6		EXPERIMENTAL SET UP	53
		3.6.1	Hot Air Generator Facility	53
		3.6.2	Gas/Hot Air Generator	54
		3.6.3	Ejector-Entraining Diffuser IRSS Device	55
		3.6.4	Principle of Operation	55
		3.6.5	Structural and Fluid Parameters for Study	56
		3.6.6	Experimental Study Methodology	56
	3.7		FABRICATED MODELS	57
	3.8		INSTRUMENTATION AND SENSORS	59
		3.8.1	Resistance Temperature Detector (RTD)	59
		3.8.2	Thermocouple Specification	60
		3.8.3	Pressure Sensor	60
		3.8.4	Infrared Imaging Camera	61
	3.9		CONCLUSION	63
CHAPTER 4			**MATHEMATICAL MODELLING AND DIMENSIONAL DESIGN**	**64-79**
	4.1		INTRODUCTION	64
	4.2		TYPICAL ARRANGEMENT OF EXHAUST ON SHIPS	64
	4.3		INPUT PARAMETERS FOR MATHEMATICAL MODEL	65
	4.4		DESIGN OF EJECTOR SUBSYSTEM	66
		4.4.1	Nozzle	66
		4.4.2	Standoff Distance	66
		4.4.3	Mixing Tube	67
	4.5		DESIGN OF DIFFUSER SUBSYSTEM	68
		4.5.1	Overall Diffuser Dimensions	67

	4.5.2	Diffuser Ring Length and Slot Width	68
4.6		EJECTOR-DIFFUSER FOR SHIP APPLICATION (EDSA)	69
	4.6.1	Input Parameters for EDSA	69
	4.6.2	Calculated Parameters	70
	4.6.3	Optimal Component Dimensional Range	71
	4.6.4	Study of Nozzle Design using EDSA	71
	4.6.5	Study of Diffuser Design with EDSA	73
4.7		AERO-THERMAL PERFORMANCE MODEL (ATPM)	73
	4.7.1	Input Parameters for ATPM	74
	4.7.2	Utility of ATPM in Preliminary Evaluation of EDSA Generated Designs	75
4.8		INFERENCES FROM STUDY USING EDSA & ATPM	77
4.9		COMPETING DESIGN CONSIDERATIONS FOR INTEGRATED IRSS	78
4.10		UTILITY OF MATHEMATICAL TOOLS	78
4.11		CONCLUSION	79
CHAPTER	**5**	**NUMERICAL SIMULATION STUDIES**	**80-120**
5.1		Introduction	80
5.2		Generation of 3D Models	80
5.3		Computational Domain	80
5.4		Grid Structure	81
5.5		CFD SOLVER	83
	5.5.1	Physics Models	83
	5.5.2	Turbulence Modelling Approach	84
	5.5.3	Discretization Scheme	85
	5.5.4	Pressure-Velocity Coupling	86
	5.5.5	Near Wall Treatment	87
	5.5.6	Boundary Conditions	87
	5.5.7	Solution Convergence Study	88
5.6		FORMULATION OF CASE STUDIES	88

5.7	CASE STUDY 1- ROUND NOZZLE RN1	90
	5.7.1 Exhaust Plume	91
	5.7.2 Diffuser Ring Duct Surface Temperatures	92
	5.7.3 Pressure Profile	93
5.8	CASE STUDY 2 - ROUND NOZZLE RN2	93
	5.8.1 Exhaust Plume	94
	5.8.2 Diffuser Ring Duct Surface Temperatures	95
	5.8.3 Pressure Profile	96
5.9	CASE STUDY 3 - ROUND NOZZLE RN3	97
	5.9.1 Exhaust Plume	97
	5.9.2 Diffuser Ring Duct Surface Temperatures	98
	5.9.3 Pressure Profile	98
5.10	CASE STUDY 4 - THREE LOBED NOZZLE	99
	5.10.1 Exhaust Plume	99
	5.10.2 Diffuser Ring Duct Surface Temperatures	101
	5.10.3 Pressure Profile	101
5.11	CASE STUDY 5 - FOUR LOBED NOZZLE	102
	5.11.1 Exhaust Plume	102
	5.11.2 Diffuser Ring Duct Surface Temperatures	103
	5.11.3 Pressure Profile	104
5.12	CASE STUDY 6 - SIX LOBED NOZZLE (6LN)	104
	5.12.1 Exhaust Plume	105
	5.12.2 Diffuser Ring Duct Surface Temperatures	106
	5.12.3 Pressure Profile	107
5.13	CASE STUDY 7 - ELLIPTICAL NOZZLE (EN)	107
	5.13.1 Exhaust Plume	108
	5.13.2 Diffuser Ring Temperature Profile	109
	5.13.3 Pressure Profile	109
5.14	CASE STUDY 8 -TWIN MIXING TUBE DESIGN (TMT)	110
	5.14.1 Exhaust Plume	110
	5.14.2 Diffuser Duct Surface Temperature	111

5.14.3 Pressure Profile 112

5.15 CASE STUDY 9 - TWO RING DIFFUSER WITH ONE
RING GAP (1RG) 112

5.15.1 Plume Temperature Profile 113

5.15.2 Diffuser Ring Temperature Profile 114

5.15.3 Pressure Profile 114

5.16 CASE STUDY10 - TWO RING DIFFUSER WITH NO
OVERLAP (0RG) 115

5.16.1 Plume Temperature Profile at Diffuser Exit 115

5.16.2 Diffuser Ring Temperature Profiles 116

5.16.3 Pressure Profile 116

5.17 CASE STUDY11- DIFFUSER WITH OVERLAP OF
2 RING GAP (2RG) 117

5.17.1 Plume Temperature Profile 117

5.17.2 Diffuser Ring Temperature profiles 117

5.17.3 Pressure Profile 118

5.18 CASE STUDY 12 - EXHAUST STACK WITHOUT IRSS 118

5.18.1 Exhaust Plume, Duct and Pressure Profiles 119

5.19 CONCLUSION 120

CHAPTER 6 EXPERIMENTAL STUDIES 121-140

6.1 INTRODUCTION 121

6.2 CASE STUDIES FOR EXPERIMENTAL VALIDATION 121

6.3 CASE STUDY 4 - IRSS WITH ROUND NOZZLE AND 4
RING DIFFUSER 122

6.3.1 Experimental Setup 122

6.3.2 Pressure Management 122

6.3.3 Exhaust Plume Temperature Measurement 124

6.3.4 Exhaust Duct Temperature Measurement 125

6.3.5 Tertiary Air through Diffuser Rings 127

6.3.6 Remarks 128

6.4   CASE STUDY 1- IRSS WITH ROUND NOZZLE (RN1)   128

  6.4.1   Experimental Setup   128

  6.4.2   Pressure Measurement   129

  6.4.3   Exhaust Plume Temperature Measurement   129

  6.4.4   Exhaust Duct Temperature Measurement   130

  6.4.5   Tertiary Air through Diffuser Rings   131

  6.4.6   Remarks   132

6.5   CASE STUDY-8 IRSS WITH TWIN MIXING TUBE

  & 2 RING DIFFUSER   132

  6.5.1   Experimental Setup   132

  6.5.2   Pressure Measurement   133

  6.5.3   Exhaust Plume Temperature Measurement   134

  6.5.4   Exhaust Duct Temperature Measurement   135

  6.5.5   Tertiary Air through Diffuser Rings   136

  6.5.6   Remarks   136

6.6   CASE STUDY- 9 & 10-DIFFUSER RING OVERLAP

  STUDY   136

  6.6.1   Experimental Setup   136

  6.6.2   Tertiary Air through Diffuser Rings   137

  6.6.3   Exhaust Plume Temperature Measurement   138

  6.6.4   Exhaust Duct Temperature Measurement   139

  6.6.5   Remarks   139

6.7   CONCLUSION   140

CHAPTER   7   RESULTS AND DISCUSSIONS   141-164

7.1   INTRODUCTION   141

7.2   GROUPING OF CASE STUDIES   141

7.3   NOZZLE DESIGN - CASE STUDIES 1 TO 7   142

	7.3.1	Primary Mass Flow Rate	143
	7.3.2	Pumping Ratio	144
	7.3.3	Exhaust Duct Temperature	147
	7.3.4	Static Pressure Profile	148
	7.3.5	Exhaust Plume Temperature at IRSS Exit	151
7.4		MIXING TUBE DESIGN STUDY- CASE STUDY 9&10	153
	7.4.1	Pumping Ratio	154
	7.4.2	Mixing Tube Thermal Profile	155
	7.4.3	Uniformity of Exhaust Plume Profile at IRSS Exit	157
7.5		DIFFUSER DESIGN STUDY-CASE STUDY 2, 10, 11 &12	158
	7.5.1	Pumping Ratios for Different Diffuser Designs	158
	7.5.2	Ring Temperatures for Different Diffuser Designs	159
	7.5.3	Exhaust Plume Temperature at Diffuser Exit	161
7.6		INTEGRATED INFRARED SIGNATURE SUPPRESSION DESIGN	161
	7.6.1	Exhaust Duct Temperatures	162
	7.6.2	Exhaust Plume Temperature	163
	7.6.3	Back Pressure	163
	7.6.4	Percentage Reduction in Missile Lock-on Range	164
7.7		CONCLUSION	
CHAPTER 8		**CONCLUSIONS AND FUTURE SCOPE**	**165-168**
	8.1	INTRODUCTION	165
	8.2	CONCLUSIONS	165
	8.3	FUTURE SCOPE OF WORK	168
	8.4	CONCLUSION	168
REFERENCES			**169-176**

LIST OF FIGURES

Page No

Figure: 1.1	Naval Ship Signature Spectrum	3
Figure: 1.2	Electromagnetic Spectrum	5
Figure: 1.3	Thermal Radiation Spectrum	6
Figure: 1.4	Atmospheric Transmittance at Sea level	8
Figure: 1.5	Major Sources of IR Signature on Ships	9
Figure: 1.6	Spectral Radiant Intensity of IRSS Duct Surfaces (Grey body)	10
Figure: 1.7	Spectral Emittance of Typical GT Exhaust Plume	11
Figure: 1.8	Typical Installation of IRSS Device in Ship Funnel	13
Figure: 2.1	Early IRSS –Multi Nozzle Ejector	21
Figure: 2.2	Cylindrical and Conical IRS Devices from Literature	31
Figure: 2.3	Entraining Diffusers from Literature	36
Figure: 2.4	Shipboard Requirements for Ejector-Diffuser IRSS Systems	37
Figure: 3.1	Workflow in Numerical Simulation using CFD	45
Figure: 3.2	Flowchart depicting Generation of 3D Model in CATIA	44
Figure: 3.3	Creation of a domain around IRSS System	46
Figure: 3.4	Mesh Models in STAR CCM+	48
Figure: 3.5	Selection of Optional Mesh Models with Core Mesh	50
Figure: 3.6	Flowchart of Mesh Generation in Simcenter STAR CCM+	51
Figure: 3.7	Procedure to Select Physics Models	52
Figure: 3.8	Experimental Setup of Integrated Ejector-Diffuser IRSS Device	54

Figure: 3.9   Principle of Operation of Ejector- diffuser IRSS Device   56

Figure: 3.10   Nozzles of different shapes and Inlet-Exit Area Ratios   58

Figure: 3.11   Single and Twin Mixing Tube Configurations   58

Figure: 3.12   Four Ring and Two Ring Diffuser Configurations   58

Figure: 3.13   Location of RTDs and Thermocouples on IRSS   59

Figure: 3.14   RTDs for Exhaust Diffuser Ring Temperatures   60

Figure: 3.15   Thermocouple for Exhaust Plume Temperature Measurement   60

Figure: 3.16   S and L-Type Pilot Tubes for Pressure Measurements   61

Figure:3.17   Thermal Imaging Camera FLIR T 1020 used in study   63

Figure: 4.1   Exhaust Duct Arrangement onboard Ships   65

Figure: 4.2   View Angle Protection for Infrared Suppression   68

Figure: 4.3   Effect of Nozzle Inlet to Exit Ratio on Back Pressure   72

Figure: 4.4   Effect of View Angle on Diffuser & Mixing Tube Design   73

Figure: 4.5   Preliminary Performance Assessment with ATP Model   75

Figure: 5.1   Computational Domain of Passive IRSS System   81

Figure: 5.2   Progressive Mesh used in the study   82

Figure: 5.3   Physics Models selection in Simcenter STAR CCM+ software   84

Figure: 5.4   Physics Models Selected for CFD Analysis of IRSS System   84

Figure: 5.5   Boundary Conditions for IRSS System   87

Figure:5.6   Graph of Residuals for Monitoring Convergence   88

Figure: 5.7   Dimensional Design and 3 D Model- Case Study-1   90

Figure: 5.8   Thermal Profile- Case Study-1   91

Figure: 5.9   Velocity Profile- Case Study-1   92

Figure: 5.10   Diffuser Ring Wall Surface Temperatures –Case Study 1   93

Figure:5.11   Static Pressure Profile at Nozzle-Case Study 1   94

Figure: 5.12   Dimensional Design and 3D Model of Round Nozzle (RN2)   94

Figure: 5.13   Thermal and Velocity Profile Along IRSS- Case Study 2   95

Figure: 5.14   Thermal & Velocity Profiles at Device Exit- Case Study 2   95

Figure: 5.15   Diffuser Ring Wall Surface Temperature for Case Study 2   96

Figure: 5.16   Static Pressure profile in IRSS for Case Study 2   96

Figure: 5.17   Thermal and Velocity Profile along IRSS- Case Study 3   97

Figure: 5.18   Thermal and Velocity Profiles at Device Exit- Case Study 3   98

Figure: 5.19   Diffuser Ring Wall Temperatures Case Study 3   98

Figure: 5.20   Static Pressure Profile at Nozzle Inlet- Case Study 3   99

Figure: 5.21   Dimensional Design and 3 D M Case Study 4   100

Figure: 5.22   Thermal and Velocity Profile- Case Study 4   100

Figure: 5.23   Thermal and Velocity Profile at Exit -Case Study 4   100

Figure: 5.24   Diffuser Ring Wall Surface Temperature- Case Study 4   101

Figure: 5.25   Static Pressure Profile at Exit and along IRSS- Case Study 4   102

Figure: 5.26   Dimensional Design and 3D Model- Case Study 5   102

Figure: 5.27   Thermal and Velocity Profile- Case Study 5   103

Figure: 5.28   Diffuser Ring Wall Surface Temperatures for Case Study 5   104

Figure: 5.29   Static Pressure Profile at Exit and along IRSS for Case Study- 5   104

Figure: 5.30   Dimensional Design and 3 D Model- Case Study-6   105

Figure: 5.31   Thermal and Velocity Profile- Case Study 6   105

Figure: 5.32   Thermal and Velocity Profile at Funnel exit for Case Study 6   106

Figure: 5.33   Diffuser Ring Wall Surface Temperatures for Case Study 6   106

Figure: 5.34   Static Pressure Profile at Exit and along IRSS for Case Study 6   107

Figure: 5.35   Dimensional Design and 3 D Model- Case Study 7   107

Figure: 5.36   Thermal and Velocity Profile- Case Study 7   108

Figure: 5.37   Thermal and Velocity Profile at Device Exit- Case Study 7   108

Figure: 5.38   Diffuser Ring Wall Surface Temperatures- Case Study 7   109

Figure: 5.39   Static Pressure Profile at Exit and along IRSS- Case Study 7   109

Figure: 5.40   Dimensional Design and 3 D Model- Case Study -8   110

Figure: 5.41   Thermal and Velocity Profile along IRSS- Case Study 8   111

Figure: 5.42   Thermal and Velocity Profile at Device Exit- Case Study 8   111

Figure: 5.43   Diffuser Ring Wall Surface Temperature- Case Study 8   112

Figure: 5.44   Static Pressure Profile at Exit and along IRSS- Case Study 8   112

Figure: 5.45   Thermal Profile at Diffuser Exit and along IRSS- Case Study 9   113

Figure: 5.46   Thermal Profile at Mixing Tube and Diffuser Exit- Case Study 9   113

Figure: 5.47   Thermal Profile of Bottom and Top Rings-Case Study 9   114

Figure: 5.48   Pressure Profile at Nozzle Inlet and Diffuser Inlet- Case Study 9   114

Figure: 5.49   Dimensional Drawing & 3 Model of Diffuser with No overlap
                - Case study 10   115

Figure: 5.50   Thermal Profile at Mixing Tube and Diffuser Exit- Case Study 10   116

Figure: 5.51   Thermal Profile of Bottom and Top Rings- Case Study 10   116

Figure: 5.52   Pressure Profile at Nozzle Inlet and along IRSS for Case 9   117

Figure: 5.53   Thermal Profile at Diffuser Exit and along IRSS- Case Study 11   117

Figure: 5.54   Thermal Profile of Bottom and Top Rings- Case Study 11   118

Figure: 5.55   Pressure Profile at Nozzle Inlet and along IRSS- Case Study 11   118

Figure: 5.56   Thermal and Velocity Profile of Exhaust Plume- Case Study 12   119

Figure: 5.57   Thermal, Velocity & Pressure at Duct Exit- Case Study 12   120

Figure: 6.1   Case Study 2-IRSS with Round Nozzle   122

Figure: 6.2   Measuring Points at Nozzle & Diffuser for Pressure & Temperature 123

Figure: 6.3   Temperature Profile at Nozzle Exit-Case Study 2   124

Figure: 6.4   Temperature Profile at Diffuser Exit- Case Study-2   124

Figure: 6.5   Temperature Profile of Mixing Tube-Case Study-2   125

Figure: 6.6   Temperature Profile of Diffuser Rings-Case Study-2   125

Figure: 6.7   Temperature Profiles of Individual Diffuser Rings-Case Study-2   127

Figure: 6.8   Case Study 1- IRSS with Cylindrical Nozzle (RN1)   128

Figure: 6.9   Temperature Profile at Nozzle Exit- Case Study 3   130

Figure: 6.10   Temperature Profile at Diffuser Exit-Case Study 3   130

Figure: 6.11   Temperature Profile of Mixing Tube-Case Study 3   131

Figure: 6.12   Temperature Profile of Diffuser Rings- Case Study 3   131

Figure: 6.13   Case Study 8-IRSS with Twin Mixing Tube and 2 Ring Diffuser   133

Figure: 6.14   Temperature Profile at Bottom Mixing Tuber Exit- Case Study 8   134

Figure: 6.15   Temperature Profile at Two Ring Diffuser Exist- Case Study 8   134

Figure: 6.16   Temperature Profile of Bottom & Top (MT 2) Mixing Tubes   135

Figure: 6.17   Temperature Profile of Two Ring Diffuser- Case Study 8   135

Figure: 6.18   One Ring Gap Overlap   137

Figure: 6.19   Nil Overlap Diffuser   137

Figure: 6.20   Temperature Profile for Diffuser (a) with (b) without overlap   138

Figure: 6.21   Temperature Profile of Bottom & Top Rings with overlap   139

Figure: 6.22   Temperature Profile of Bottom & Top Rings without overlap   140

Figure: 7.1   Primary Mass Flow at Diffuser Exit for Different Nozzle Shapes   144

Figure: 7.2   Simulated and Measured Pumping Ratios for different Nozzles   145

Figure: 7.3   Velocity at MT Exit for Round Nozzle   146

Figure: 7.4   Velocity Profile at Diffuser Exit for Round Nozzle   146

Figure: 7.5   Simulated Wall Temp. of Diffuser Rings for Case Studies 2 to 6   147

Figure: 7.6   Simulated Thermal Profile, Measured IR Image Plot- Case Study 2   148

Figure: 7.7   Simulated Thermal Profile, Measured IR Image Plot- Case Study 3   148

Figure: 7.8   Pressure Profile along IRSS for Round, Lobed & Elliptical Nozzles   149

Figure: 7.9   Static Pressure Profile at Nozzle Inlet for Round & Lobed Nozzle   149

Figure: 7.10   Effect of Nozzle Shape on Pressure Profile at Inlet to IRSS   150

Figure: 7.11   Simulated Temperature at IRSS Exit for Case Studies 2, 3, 4, 5 & 6   151

Figure: 7.12   Simulated & Measured Temperature Profiles at IRSS Exist   152

Figure: 7.13   Average Plume Temperature at IRSS Exit- Simulated & Measured   152

Figure: 7.14   Plume Velocity along IRSS- RN1 & RN2   153

Figure: 7.15   Pumping Ratio of Single & Twin MTs- Case Studies 9 & 10   154

Figure: 7.16   Thermal Profile of Single & Twin MTs through simulation   155

Figure: 7.17   Temperature along Length of Twin Mixing Tubes (Top Half)   156

Figure: 7.18   Temperature along Length of Single Mixing tube   156

Figure: 7.19   Temperature at Exit Plane for (a) Single & (b) Twin Mixing Tubes   157

Figure: 7.20   Diffuser Design with Varying Number of Rings & Overlap   158

Figure: 7.21   Experimental models of case studies 2, 9 and 10 for diffuser design   159

Figure: 7.22   Simulated Thermal Profile, Measured IR Image Plot-case study 2   160

Figure: 7.23   Measured IR image Plots of Diffuser Top Ring Temperatures   160

Figure: 7.24   Performance study of (a) non-IRSS (b) Passive IRSS   162

Figure: 7.25   Simulated thermal profile of exhaust without IRSS and with IRSS   162

Figure: 7.26   Thermal profiles of Plume without IRSS and with IRSS   163

LIST OF TABLES

Page No

Table 2.1	Typical Fluid and Structural Parameters of Ship Propulsion	38
Table 3.1	Specification of Hot Air Generator	54
Table 3.2	Input Structural and Fluid Parameters for Study	56
Table 3.3	Specifications of Thermal Imaging Camera used in Study	62
Table 4.1:	Input Fluid Parameters of EDSA	69
Table 4.2:	Fluid and Structural Parameters Calculated by EDSA	70
Table 4.3:	Range of Output Dimensions Provided by EDSA	71
Table 4.4:	Nozzle Design Study for EDSA	72
Table 4.5:	Input Parameters for ATPM	74
Table 4.6:	Inferences from Study using EDSA and ATPM	76
Table 4.7:	Critical Parameters of Ejector Diffuser IRSS Device	77
Table 5.1:	Details of Mesh for Computational Domain of IRSS	82
Table 5.2:	Discretization schemes available for RANS models in FLUENT	85
Table 5.3:	Formulation of Case Studies for Numerical Simulation	89
Table 6.1:	Measurement of Static and Dynamic Pressure –Case Study 2	123
Table 6.2:	Measurement of Tertiary Air Induced –Case Study 2	127
Table 6.3:	Measurement of Static and Dynamic Pressure –Case Study 3	129
Table 6.4:	Measurement of Tertiary Air Induced –Case Study 3	132
Table 6.5:	Measurement of Static and Dynamic Pressure –Case Study 8	133
Table 6.6:	Measurement of Tertiary Air Induced –Case Study 8	136
Table 6.7:	Measurement of Tertiary Air Induced –Case Study 9 & 10	138
Table 7.1	Grouping of Case Studies	142
Table 7.2	Parameters for Design Study of Nozzle Subsystem	143
Table 7.3	Ring Surface Average Temperature for Diffuser Design Study	160

NOMENCLATURE

A	Area
A_{ni}	Area at nozzle inlet
A_{ne}	Area at nozzle exit
A_{mt}	Area of mixing tube
A_{de}	Area of diffuser exit
A_{jet}	Area of the jet (exhaust gas) at nozzle exit
C_p	Specific heat at constant pressure
D_{ni}	Diameter at nozzle inlet
D_{ne}	Diameter at nozzle exit
D_{mt}	Diameter of mixing tube
D_{de}	Diameter at the Diffuser exit
dW	Small amount of work
dt	Small interval of time
K-ε	Turbulence model in CFD
L	Total length of IRSS device
LOR	Lock on Range of the 1st model IRSS
$\dot{m}$	Mass Flow Rate
$\dot{m}_{ni}$	Mass flow rate of Primary fluid at nozzle inlet
$\dot{m}_{mti}$	Mass flowrate of secondary fluid induced at mixing tube inlet
$\dot{m}_{de}$	Total Mass flowrate at the diffuser exit
Ma	Mach number
N_b	Radiance of background
N_t	Radiance of Target
N_{t0}	Radiance of target without IRSS (Non-stealth ship)

N_{t1}	Radiance of Target with typical IRSS device (Stealth ship)
P_1	Pressure at the nozzle exit
P_2	Pressure at the nozzle exit
P_b	Back Pressure due to IRSS on engine
P_{ni}	Pressure of Primary fluid at nozzle inlet
Pr	Prandtl Number
Q	Heat capacity
Re	Reynolds Number
R_t	Radius of the top ring
R_b	Radius of the bottom ring
S	Standoff distance
T_{amb}	Temperature of ambient air
T_{de}	Temperature of Plume at the exit of diffuser at location-1
T_e	Bulk mean temperature of the plume at the exit of IRSS
T_{ni}	Temperature of Primary fluid at nozzle inlet
T_{r1}	Wall Temperature of the Diffuser Ring-1
T_{r2}	Wall Temperature of the Diffuser Ring-2
T_{r3}	Wall Temperature of the Diffuser Ring-3
T_{r4}	Wall Temperature of the Diffuser Ring-4
V_{ni} (V_1)	Velocity of gas at nozzle inlet
V_{ne} (V_2)	Velocity of gas at nozzle exit
Vs	Velocity of the Secondary Fluid at standoff distance
V_{r1}	Velocity of the Tertiary Fluid at the entry of Ring-1
V_{r2}	Velocity of the Tertiary Fluid at the entry of Ring-2
V_{r3}	Velocity of the Tertiary Fluid at the entry of Ring-3
V_{r4}	Velocity of the Tertiary Fluid at the entry of Ring-4
Z	Height (in Bernoulli's equation)

Greek symbols

α	Nozzle convergence angle
β	Diffuser half angle
ε	Emissivity
$^\circ$	Degree
θ	View Angle Protection
Φ_s	Pumping ratio of secondary air through standoff
Φ_t	Pumping ratio of tertiary air through diffuser rings
Φ_T	Pumping ratio of total IRSS
σ^2	Variance of Exhaust Gas Temperature
γ	Ratio of Specific heats
ρ	Density

List of Acronyms & Abbreviations

Amb	Ambient
ATPM	Aero-Thermal Performance Model
BLISS	Boundary Layer Infrared Signature Suppression
CFD	Computational Fluid Dunamics
CIE	International Commission on Illumination
DE	Diesel Engine
DRES	Defence Research Establishment Suffield, Canada
EDSA	Ejector-Diffuser for Ship Applications

Fig.	Figure
GT	Gas Turbine
HAG	Hot Air Generator
IR	Infrared
IRSS	Infrared Signature Suppression
PR	Pumping ratio
LOR	Lock on Range
LWIR	Long wave infrared
MT	Mixing Tube
MWIR	Medium wave infrared
NBC	Nuclear Chemical Biological
NSTL	Naval Science and Technology |Laboratory, Visakhapatnam
RTD	Resistance Temperature Detector
SMT	Single Mixing Tube
TC	Thermocouple
TMT	Twin Mixing Tube
UV	Ultra Violet

CHAPTER 1

INTRODUCTION

1.1 BACKGROUND

In 450 BCE, Sun Tzu wrote in his treatise, 'The Art of War', that *All warfare is based on deception*. An analogous view was presented in the Arthashastra composed by Chankya around 320 BCE, in which it is stated that *secrecy and dissimulation are essential to warfare*. In the Soviet Union, the term *maskirovka* referred to any strategy or tactic in which deception was central and included practices such as use of camouflaged or fake military positions during the Russian Civil wars of 1920s. In modern warfare, the role of stealth became important with advent of Radars. The act of deceiving the enemy by a reduced giveaway signature is at the core of stealth philosophy.

In the vast blue ocean, ships of different navies operate far from each other and are generally not in visual range. However, a ship can be uniquely identified based on a number of emissions that are characteristic to each and are called signatures. These include acoustic signature from machinery and flow, heat or infrared signature from exhaust, radar cross section from hull surfaces, magnetic signature from hull material, electric signature from corrosion protection systems etc. These signatures are used by surveillance and tracking sensors for attacking ships in general and warships, in particular. Survivability at sea requires management of all platform signatures to produce a net effect of reduced vulnerability.

Anti-ship missiles pose the greatest threat to ships at sea. Prior to the Second World War, studies were carried out in parallel on the use of radar and infrared seekers in missiles for detection, identification and homing. For the materials available in the 1940s, radar seekers gave better results and hence, have become the standard fit. Over the years, advances in material sciences have improved the capabilities of infrared seekers and today, missiles have several guidance/seeker options that include radar, electro-optical camera and infrared imaging. Infrared homing is preferred for terminal guidance because of its inherent immunity to jamming techniques. Developments in the field of

1

missile homing led to counter measure developments and need for installing infrared suppression systems in ship exhaust systems began receiving attention since the late 1980s. In his study on ship survivability, Kok (2012) estimated that use of infrared signature suppression devices reduced the probability of detection from 85% to 50% for a frigate sized vessel.

Studies in the field proposed optimal dimensions of air-air ejectors, louvered funnel, and entraining diffusers to facilitate entrainment of large amounts of ambient air to cool the exhaust plumes. Aim of the studies was predominantly for academic research and thus, has not considered practical shipboard space constraints, engine back pressure limitations and marine engine exhaust plume parameters while proposing the optimal dimensions. Further different designs are compared based on their capability for air entrainment, pressure recovery etc., rather than infrared signature reduction. The research work proposes to augment existing knowledge by including shipboard constraints as the basis for design of infrared suppression devices.

1.2 NAVAL SHIP SIGNATURES

A stealthy naval platform remains undetected and therefore can be deployed closer to the target to fulfill its mission with lesser risk of damage and associated causalities. Signatures used for ship detection include both underwater and above water as shown in Fig. 1.1. Signatures used by underwater weapons like torpedoes and mines are acoustics, magnetics, wake homing and extremely low frequency emissions (ELFE). Anti-ship missiles use above water signatures that include laser cross section (LCS), radar cross section (RCS), air borne noise and infrared (IR). A brief description of the source generating the signature and available mitigation methods are enumerated in succeeding paragraphs.

1.2.1 Acoustic Signature

Surface warships have strong sources of underwater noise which can easily be detected by passive and active sonar from very long distances up to hundreds of km away. Reduction of acoustic signature is achieved by using vibration isolation mounts for machinery, epicyclical gearboxes, shrouded or contra rotating propellers, fluid couplings etc. Active Noise Cancellers essentially consist of a sensor for detecting the presence of sound, inverter for phase inversion and actuators for retransmission of the sound with an inverted phase.

Fig. 1.1 Naval Ship Signature Spectrum

1.2.2   Magnetic Signature

One of the most effective weapons against ships in littoral waters is the naval mine. Modern influence mines often sense the static (or dc) magnetic fields of passing ships to be used as triggers. Permanent and induced magnetic fields associated with ship's steel hull are reduced by de-perming and degaussing respectively.

1.2.3   Wake Homing Signature

Hydrodynamic wake is generated by surface ships as well as submarines. Ship wake signatures have several distinct features such as a dark trailing centerline region, bright-line images aligned at some angle to the ships path, and sometimes the Kelvin wake pattern. Developments in airborne and satellite borne remote sensors have enabled

3

not only detection but also classification. Studies on wake cancellation hull forms and fluid clock technology are at a nascent stage of research.

1.2.4 Extremely Low Frequency Emissions (ELFE)

Galvanic currents flowing in the hull generate underwater electrical potentials. Under certain conditions, this can cause extremely low frequency emissions (ELFE) to be radiated into the water. Detection of the same is a recent capability. Impressed Current Cathodic Protection (ICCP) systems, Zinc anodes on hull plates at waterline are some of the sources. Passive and active shaft grounding techniques are used to suppress the signature.

1.2.5 Radar Cross Section (RCS)

Active Radar Homing is the preferred mode for most missiles and hence, management of RCS signature is of critical importance. Radar cross section depends on hull form, upper deck appendages and construction material. Hull shaping, Radar Absorbent Paints (RAP) and Radar Absorbent Materials (RAM) are known to achieve low RCS signature.

1.2.6 Laser Cross Section

The ship signature measured using a LIDAR (Light detection and ranging) in place of radio waves is called lase Cross Section and is distinct from radar cross section. The shorter wavelengths of light used by LiDAR enable detection of details over smaller cross section, thus making the ship highly susceptible for detection. Studies on measurement of laser cross section and techniques for mitigation are at a nascent stage.

1.2.7 Air Borne Noise Signature

Sources generating air borne noise include machinery casings, propulsion, intakes and exhausts, piping and air-conditioning (HVAC) systems. When airborne sound propagates in a free field, it gradually attenuates over distance. However, when it meets a solid object, such as a steel plate, the attenuation increases significantly (as much as 40

4

dB). The joiner panel, Sound absorptive material such as fiberglass applied to the partition can also decrease the transmission of noise.

1.2.8 Thermal / Infrared (IR) Signature

IR detectors in missile homing head sense the radiant energy emitted by objects within its search zone. It is proportional to fourth power of temperature (T^4) and therefore, areas with high temperature such as exhaust funnels are seen as hotspots from longer distances and are used for terminal guidance of the missile. Infrared signature reduction of ships is realized through methods such as temperature reduction, line of sight masking and emissivity control.

1.3 ELECTROMAGNETIC RADIATION

Electromagnetic spectrum covers a wide range of wavelengths and photon energies as shown in Fig. 1.2. Infrared radiation has wavelengths longer than visible light and hence, is not visible. Even though the peak of Sun's emission spectrum is in visible range (0.5 μm) as shown in Fig. 1.3, over 50% of Sun's energy reaching the earth is in the form of infrared. Infrared radiation is therefore loosely referred to as thermal radiation. But thermal radiation (0.1 to 1000 μm) includes UV radiation (0.1 to 0.4 μm), visible light (0.4 to 0.76 μm) and Infrared (0.76 to 1000 μm). The balance between infrared radiation absorbed by earth during the day and emitted at night has a significant impact on climate.

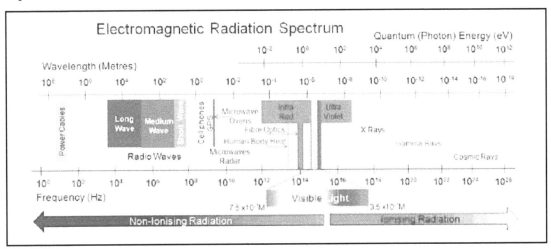

Fig. 1.2 Electromagnetic Spectrum

5

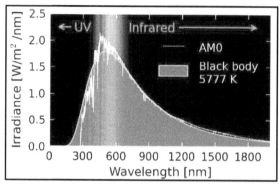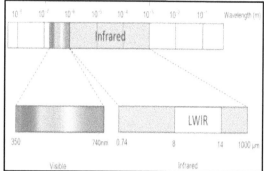

Fig. 1.3 Thermal Radiation Spectrum

1.4 INFRARED (IR) RADIATION

Sir William Herschel carried out over 200 experimental observations as part of his efforts to determine the suitable filter to observe the Sun and in March 1800, noted, "radiant heat will at least partly, if not chiefly, consist, if I may be permitted the expression, of invisible light; that is to say, of rays coming from the sun, that have such a momentum as to be unfit for vision." Infrared light has a longer wavelength (and so a lower frequency) than that of the red light visible to humans. The term 'infrared' was coined in 1880, which literally means '*below red*' because the Latin word '*infra*' means '*below*'. Broadly, it includes electromagnetic wavelengths in the range of 0.700 µm to 1000 µm.

Infrared radiation is emitted or absorbed by molecules when changing rotational-vibrational movements. It excites vibrational modes in a molecule through a change in the dipole moment, making it a useful frequency range for study of these energy states for molecules of the proper symmetry. Infrared radiation is produced by objects whose temperature is above 10°K. Black-body radiation from objects near room temperature is almost all at infrared wavelengths.

1.4.1 Sub Division of Infrared Wavebands

The infrared radiation band extending from 0.7 µm to 1000 µm is sub divided into various regions. Range of infrared wavebands for each region are not standardized and vary based on type of detector technology used for gathering infrared light for the application. A few of the sub divisions include ISO 20473 scheme for optics and photonics, CIE Division scheme for optical, visual and metrological aspects of the communication, astronomy division scheme for celestial observations, Sensor Response Division scheme for infrared detectors and Telecommunication bands for optical communication. For infrared signature purposes, the waveband from 3-5 μm is considered medium wave (MWIR) and the waveband from 7-14 μm is considered long wave (LWIR).

1.4.2 Atmospheric Transmission of Infrared Radiation

Infrared based detection is a passive technology in that the detector can only receive the infrared radiation emitted by the source. But the detected radiation is not the same as at the source because it has to pass through earth's atmosphere and as it travels, it is absorbed in selective wavebands by the atmospheric plumees such as carbon dioxide (2.7 μm ,4.3 μm and 15 μm), water vapour (based on thickness of perceptible water), industrial plumees such as carbon monoxide, ammonia, hydrogen sulfide and is scattered away from the line of sight by small particles suspended in the atmosphere such as salt from ocean spray, fine dust, carbon soot from combustion and water droplets in fog, rain, snow, clouds etc. The process by which radiant flux is attenuated in passing through the atmosphere is called extinction. The transmittance (τ) of infrared radiation through the atmosphere can be expressed as $\tau=e^{-\sigma x}$, where x is the path length and σ is called the extinction coefficient. The extinction coefficient is due to absorption and scattering.

In the wave lengths associated with IR radiation, the phenomenon of absorption affects transmittance far more than the scattering process. Transmittance in the atmosphere is different at sea level compared to high altitude regions. The transmittance of the atmosphere with respect to infrared radiation is shown in Fig 1.4. The molecules responsible for spectral absorption are shown in the figure. It can be seen that the most abundant plumes in the atmosphere viz., nitrogen, oxygen, and argon, produce no absorption in the infrared as they all have symmetrical molecules. Most of the atmospheric absorption at sea level is caused by water vapor. The two wide regions of high transmission viz., 3-5 μm (MWIR) and 7-14 μm (LWIR) wavebands are called atmospheric windows and are of interest for shipboard applications.

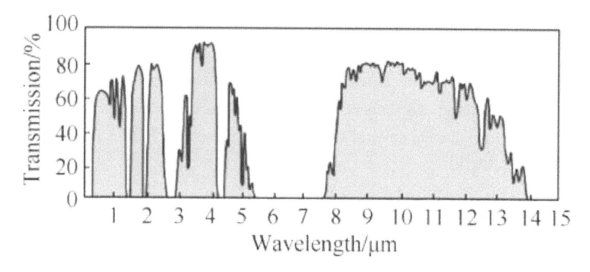

Fig. 1.4 Atmospheric Transmittance at Sea level

(Courtesy: https://commons.wikimedia.org/wiki/File:Atmosfaerisk_spredning. Png)

1.4.3 Applications of Infrared Radiation based Technologies

The earliest application of infrared technology was in the military given that it can see beyond visual. As early as 1910, work commenced on using the selenium detectors to design infrared search devices that looked for thermal signatures. Hudson (1969) enumerated various applications of Infrared technologies that included search, track & range, radiometry, spectro-radiometry, thermal imaging, and reflected flux and their utility in military, industrial, medical and scientific fields. Over last five decades, the use of infrared technologies has become even wider in fields such as condition monitoring, meteorology, ecology, forestry, agriculture, chemical and many other disciplines. The subject of the thesis being suppression of infrared signature associated with a source, wider discussion on applications that use infrared radiation for detection, tracking and homing is not included.

1.5 SOURCE OF SHIP INFRARED SIGNATURE

The radiant emittance from heat sources onboard a naval platform is specific to the class of ships and is termed as its infrared signature. The ships have a number of heat sources originating from equipment, machinery and systems required to perform its design function viz., *'to float, to move and to fight.'* Based on radiant emittance, three predominant sources enable detection of a marine platform from long range viz., exhaust plume, exhaust duct surfaces and hull surfaces. Thermal image of a typical ship is shown in Fig. 1.5. The location and contribution of the three sources is discussed.

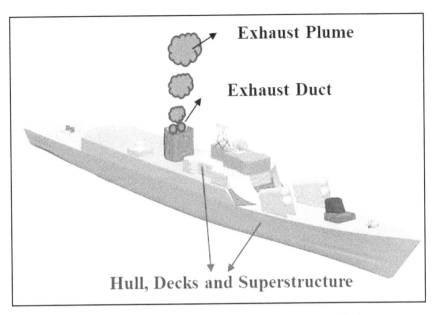

Fig 1.5 Major Sources of IR Signature on Ships

1.5.1 Exhaust Duct Assembly

The exhaust ducts of a marine platform carry the hot exhaust plumes from the propulsion and power generation engines and release them to the atmosphere at designated height above the main deck so that they cause minimum disturbance to the personnel working at various locations on the open decks or main masts of the ship. Typically, the ducts are lined with thermal insulators on the external side to prevent accidental burn injuries. Further, all the ducts from the propulsion and power generation equipment are enveloped within an enclosure called funnel. The hot exhaust ducts act as grey body radiators and radiate in all wavelengths within the Infrared spectrum as shown in Fig 1.6. Given that uncooled ship diesel engine plumes are in the range of 350 °C to 600 °C, the peak wavelength and bulk of the emittance lies within 3-5 μm wavebands. The duct inner walls are nearly at the same temperature as the exhaust plume and thus, form the brightest hotspot in the ship. The IR seeker in a missile homes on to the thermal radiation emitted by the inner walls of the individual exhaust ducts. The angle the missile subtends with the inner wall, referred to as *view angle*, increases as the missile advances towards the ship and the increased emittance is used for lock-on and homing.

9

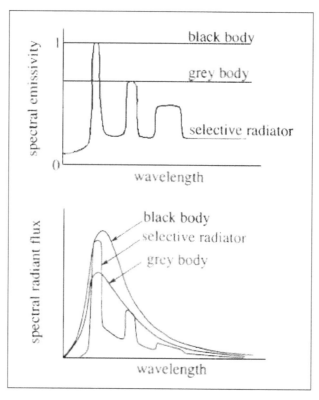

Fig 1.6 Spectral Radiant Intensity of IRSS Duct Surfaces (Grey body)

(Courtesy: http://sar.kangwon.ac.kr/etc/rs_note/rsnote/cp1/cp1-7.htm)

1.5.2 Exhaust Plume

Marine platforms use high speed diesels as fuel for propulsion and power generation equipment. Exhaust emissions from marine diesel engines comprise nitrogen, oxygen, carbon dioxide (CO_2), carbon monoxide (CO), oxides of sulphur (SOx), nitrogen oxides (NOx), hydrocarbons, water vapour and smoke. Based on their temperature, these species emit infrared radiation in specific wavelengths and hence, are called spectral radiators. Typical IR spectrum of diesel emission, for complete combustion, is shown in Fig 1.7 (Ganguly Rao, 2011). Similar to uncooled exhaust duct metal, the peak wavelength and bulk of the emittance from exhaust plume lies within 3-5 μm waveband. In fact, for complete combustion, the contribution of plume in 7-14

μm is considered negligible. The effect of incomplete combustion, which is the natural condition on marine platforms, includes the emittance due to unburnt plumes and soot. Soot is an extremely important factor because it acts as a grey body radiator at very high temperature, albeit with a very small surface area (~250nm).

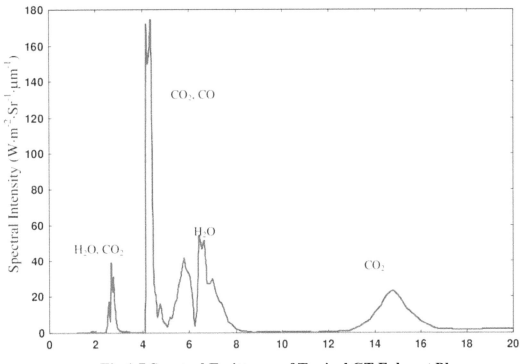

Fig 1.7 Spectral Emittance of Typical GT Exhaust Plume
(source: Ganguly Rao, 2011)

1.5.3 Hull and Associated Surfaces

Marine platforms have large hull surfaces and super structure that are heated by the Sun. The hull plate temperature is a function of thickness and emissivity. Given that the solar load during the day and night are constantly changing, the thicker plates tend to heat/ cool slower while thinner plates respond faster. The contrast radiant intensity of the large hull surfaces vis-à-vis the background sea water at ambient temperature is used by detectors in the 7-14 μm wavebands. The contribution of hull in the 3-5 μm wavebands is considerably low. However, secondary heat zones due to exhaust plume impingement on the masts, super structure and the deck surfaces, are at a far higher temperature (~ 150 °C). Exhaust ducts are sometimes routed horizontally, to avoid congestion in the funnel, and terminate at the ship side, usually, above water line. The plates attached to the exhaust duct tend to be heated up more than the adjoining plates and form a hot spot as seen in Fig. 1.6. The secondary heat zones of the hull contribute to signature in the 3-5 μm wavebands.

1.6   INFRARED SIGNATURE SUPPRESSION TECHNOLOGIES

It is known from literature that a number of technologies are employed for the suppression of infrared signature onboard ships. A brief overview of these technologies is presented below.

1.6.1 Cooling of Exhaust Plume and Exhaust Duct Surfaces using Ambient Air

The induction of large amount of ambient air to cool hot exhaust plume and associated duct surfaces through louvered openings in the exhaust ducts is proposed using two techniques. The first is by reducing pressure of exhaust plume in the duct to cause passive suction of ambient air and the second is use of blowers to force air through louvers.

Use of four different types of such systems onboard ships is known (Munro et al., 2001). Passive induction of air to cool exhaust plume and ducts is used by Eductor BLISS design IR suppressors onboard USS Spruance class ships and Eductor-Diffuser IR suppressors onboard Canada's DDH 280 destroyers. Use of forced blowers to pump in ambient air through slots is used by Cheese Grater design IR suppressors onboard Standard Frigates (SF) Batch III of Royal Netherlands Navy and DRES Ball IR suppressors onboard Canadian City class Petrol Frigates. Being of military application, details of system design and/or performance is not available. Literature is restricted to highlighting signature reduction achieved by the vessels fitted with infrared suppression systems. Typical Installation of IRSS Device in Ship Funnel is shown in Fig 1.8.

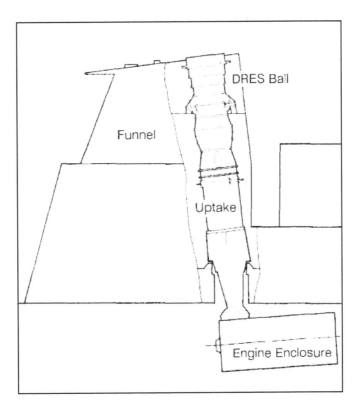

Fig 1.8 Typical Installation of IRSS Device in Ship Funnel

(Source: Birk and Vandam, 1989)

1.6.2 Cooling of Hull Surfaces

The need to cool hull surfaces to reduce IR signature due to solar loading using water curtains on ship sides and NBC wash of plates is discussed more at a conceptual level. Similarly, use of low emissivity paints as a means of reducing radiant emittance/ IR signature from secondary heating of main masts and superstructure due to exhaust plume impingement is discussed. Use of such systems onboard ships is not documented in literature.

1.6.3 Challenges in Design of IR Suppression Systems for Ship Applications

The challenges in developing practical IR suppression systems for shipboard application are as follows: -

(a) Back pressure imposed on the engine due to IR suppressor installed in exhaust system. Higher back pressure is known to increase turbine temperatures and fuel consumption for given output as also reduce engine life.

13

(b)   Additional pressure loss due to ejector is to be within engine supplier specified limits since it affects rated ship speed at full power which is a critical requirement for warships.

(c)   Availability of secondary fluid for entrainment is limited by louver area

(d)   Diffuser lengths are affected by infrared signature suppression targets

1.7   PROBLEM DEFINITION

Ejectors and diffusers are simple mechanical components that are extensively used in engineering systems. Use of an integrated ejector-diffuser device is not known to have been considered for shipboard applications. In consideration of the same, following problems are identified that need further studies to augment the available knowledge: -

(a)   Study the limitations imposed by space constraints associated with ship applications on dimensional design of ejectors and diffusers to be used for infrared suppression of ship exhaust

(b)   Examine the effect of changes in component dimensions/ shapes on the performance of an integrated ejector and entraining diffuser device with respect to infrared signature related parameters viz., exhaust plume, duct wall metal temperatures and back pressure

(c)   Experimental studies to evaluate effectiveness of alternate designs of integrated ejector-diffuser systems in suppressing infrared signature; Prediction of numerical simulation to be compared with experimental measurements.

1.8   MOTIVATION

Infrared signature suppression devices installed in gas turbine and diesel engine exhaust systems are vital to survivability of naval platforms at sea. The challenges associated with development of ejector-diffuser systems for ship applications are unique

14

and require an in depth understanding of operational constraints and requirements. Earlier research focused on simulation based optimal design of ejectors and diffusers independently. The present work intends to study the design of an integrated ejector-diffuser device specific for ship applications and is motivated by the desire to support practicing engineers involved in development of IRSS devices for naval requirements.

As part of the work, it is proposed to carryout research to address the areas identified in the problem statement that includes study of following:-

(a)   Development of suitable mathematical tools for dimensional design of an integrated ejector–diffuser and characterization of exhaust flow from infrared signature suppression perspective. Ship Builders can use the mathematical tools for preliminary assessment of ejector-diffuser designs meeting targeted IR suppression which is a mandatory requirement for funnel design at concept stage.

(b)   Using the mathematical tools, different designs of nozzles, mixing tubes and diffusers are generated and numerical simulations are carried out to study the effect of component level design variation on IR suppression capability of an integrated device. The simulations shall focus on predicting exhaust duct metal temperatures, plume temperatures, back pressure on engine to augment the existing literature on air entrainment capabilities of ejectors and pressure recovery characteristics of diffusers. The data will be of interest to product designers from industry and researchers from academia/ R&D establishments in the field of IR management.

(c)   Experimental testing of select integrated ejector-diffuser devices is carried out to validate the output of mathematical models and simulation studies. The methodology and instrumentation details will be useful to Warship Overseeing Teams involved in acceptance trials of IRSS systems prior installation on naval ships.

1.9   OBJECTIVES

Objectives of the present research are as follows: -

(i)   Carry out studies on flow parameters of exhaust gas within an integrated ejector-diffuser infrared suppressor installed in funnel onboard ships to understand effect of component level design variations on achieving targeted infrared signature suppression.

(ii)   Develop mathematical tools for preliminary dimensional design and aero-thermal performance assessment of an integrated ejector-diffuser infrared suppressor

(iii)   Carry out numerical simulations to augment available literature on ejector and diffuser designs in terms of infrared signature reduction capabilities of different designs

(iv)   Experimental tests on full scale devices to validate mathematical tools and simulation results. Collect data of exhaust plume temperatures, duct wall temperatures, static pressure and velocity of secondary air entrained for each. Infrared imaging extracted data is presented.

(v)   Identify limitations of existing methods. To enhance knowledge of full ship signature, further work is recommended

1.10 ORGANIZATION OF THE BOOK

The book is organized into eight chapters covering following broad areas: -

Chapter 1   Brief background on the importance of suppressing various ship signatures as part of naval stealth is provided. Infrared as a source of radiation and ship board signature is discussed. Known technologies for suppression of infrared signature are highlighted along with challenges in designing for ship applications. Motivation to take up the research work to overcome problem areas is presented. Broad objectives of the research are enumerated.

Chapter 2   Review of literature is carried out in four sections covering different aspects of the research area. The first section traces the developments in the field of ship infrared signature suppression technologies since 1930s. The second section reviews literature on optimal dimensional design of subsonic air-air ejectors while the third section deals with entraining diffusers. The fourth section examines the fluid and structural parameters associated with typical propulsion systems used onboard naval ships. Technology gaps are identified.

Chapter 3   Brief description of numerical simulation tool used to gain deeper insights into the complex nature of fluid flows in ship exhaust system is presented. Experimental facility available at Naval Science and Technology Laboratory, Visakhapatnam for performance assessment of IRSS systems is described. Brief description of the materials used in the current study that include hot air generator, scale models, infrared imaging cameras, sensors and instrumentation is provided.

Chapter 4   The chapter provides an overview of the shipboard arrangement of exhaust systems and input parameters affecting the design of IRSS systems. Development of two mathematical models, namely, 'Eductor-Diffuser for Ship Applications (EDSA)' for dimensional design and 'Aero-Thermal Performance Model (ATPM)' to predict performance parameters of integrated ejector-diffuser IRSS are developed.

Chapter 5   The chapter details numerical simulation of 12 case studies undertaken using Computational Fluid Dynamics software STAR CCM+. A brief overview of Pre-processing activities such as meshing, choosing suitable computational domain, grid independence study and the physics set up is provided. Performance parameters 12 case studies obtained through numerical simulation is discussed.

Chapter 6   The chapter presents experimental trials of six different IRSS systems carried out at the Diesel Engine IRSS test facility at Naval Science and

Technology Laboratory (NSTL), Visakhapatnam. The collected data of exhaust plume temperatures, duct wall temperatures, static pressure and velocity of secondary air entrained for each trial is presented. Infrared imaging is undertaken and extracted data is presented.

Chapter 7   The chapter groups the case studies based on design of key subsystem and analyses the effect of component design on the performance of integrated IRSS system in terms of pumping ratios, fluid parameters at device exit, diffuser ring thermal profile, back pressure due to IRSS and Missile Lock-On Range with respect to the baseline infrared signature associated with non-stealth ships. Studies on ship hull plate signature

Chapter 8   Salient contributions of the research work and conclusions drawn from the research work are presented. Future scope of work both in terms of enhancing available literature on passive IRSS technology as also in exploring alternate technologies for holistic management of full ship infrared signature is recommended.

CHAPTER 2

LITERATURE REVIEW

2.1 INTRODUCTION

Review of literature is carried out in four sections covering different aspects of the research area. The first section traces the developments in the field of ship infrared signature suppression technologies since 1930s. Ejector-diffuser infrared signature suppression technology is one of the effective means of IR suppression identified from the literature reviews. The second section reviews literature on optimal dimensional design of subsonic air-air ejectors for shipboard exhaust plume applications while the third section deals with entraining diffusers. The fourth section examines the fluid and structural parameters associated with typical propulsion systems used onboard naval ships. Technology gaps in the development of infrared signature suppression techniques for shipboard applications are identified.

2.2 SHIP INFRARED SIGNATURE SUPPRESSION TECHNOLOGIES

Kutzcher (1957) traced the developments on use of infrared as a source of target detection commencing early 1930s in Europe which included areas such as spectral bands of infrared radiation emitted by various targets, atmospheric attenuation, optical materials suitable for detectors, effect of background radiation on detection etc. By 1940s (Second World War), IR devices could detect and track ships from a distance of 10 to 35 km based on the temperature difference, size of target and weather conditions. However, their actual utilization was limited due to installation and system integration related issues.

Nichols (1959) examined the military applications of infrared techniques that covered missile seekers, fire control, far infrared viewers, reconnaissance and other such applications. They are predominantly for airborne systems application. The techniques included detection and tracking of surface ships by IR seekers in anti-ship missiles.

Hudson (1969) prescribed an engineering approach to designing, building and testing equipment based on infrared. In one of the most comprehensive books on infrared system engineering, he summarized the developments related to medical, industrial and military applications. He observed that infrared homing is preferred for terminal guidance of the missile because of its inherent immunity to jamming techniques.

Darling (1973) demonstrated through computer solved equations and experimental verification that a single nozzle eductor system can be used onboard naval ships to cool GT exhaust plumees. He identified the geometric parameters affecting ambient air induction. He also examined the pumping power of multiple eductor systems vis-à-vis weight increase.

In 1975, USS Spruance class ships became the first large US navy ships to use plume turbines for main propulsion. The exhaust stack was fitted with first known infrared signature suppression system onboard with the single nozzle eductor with a Boundary Layer Infrared Signature Suppression (BLISS) cap. It was based on a detailed report on design of an eductor for plume turbine propulsion ships by Hussmann in 1953.

As part of research at the Naval Postgraduate School, USA, Ellin (1977) carried forward the work of Darling by carrying out a detailed study on use of multiple nozzles for exhaust plume eductor systems. The effects of uptake Mach number, number of primary nozzles, length of primary nozzles, louver area, mixing stack inlet to primary nozzle exit area ratio on performance of eductor system are discussed in his thesis. The multiple nozzle design is implemented in GT exhaust systems onboard USS Ticonderoga (CG 47) class and USS Arleigh Burke (DDG 51) class ships [6]. The multi-nozzle eductor proposed by Ellin is shown in Fig. 2.1.

Established in 1975, W.R. Davis Engineering Limited, Canada in partnership with Defence Research Establishment Suffield (DRES), Canada, pioneered application oriented research in the field of infrared signature suppression systems for ships and aerial platforms. Majority of the literature in the 1990s owed its origin to the work done by the research establishment.

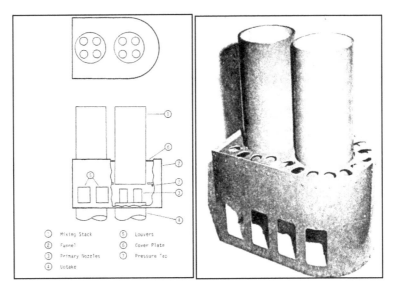

Fig 2.1 Early IRSS - Multi Nozzle Ejector

(Source: Ellin and Pucci, 1977)

In 1989, Birk and Vandam described various methods of marine plume turbine infrared suppression that included self-entraining diffusers (SED) and film cooled cylinders (FCC). The paper provides a holistic view of aero-thermal parameters that need to be considered while designing shipboard IRSS systems. However, quantitative dimensional details or performance values were not included in view of the confidential nature of developments.

In 1989, Birk and Davis reported that the research at DRES, Canada led to the development of two different designs of infrared suppression systems for shipboard applications. The first one, aptly named DRES Ball, provides optical blockage of hot exhaust of ship stack, albeit with higher back pressure. The second one based on Eductor –Diffuser design provided reasonable suppression at lower back pressure. The paper discusses details of hardware developed and issues such as back pressure on engine, noise, effect on CG, overall weight etc., related to installation onboard Canadian City class Petrol Frigates.

In 1992, Birk and Vandam presented the first and only available literature on sea trials of the DRES Ball IRSS system installed onboard the Canadian City class patrol frigate. Performance of the system was analyzed for metal temperatures of view angle protection, pumping coefficient of nozzle and plume temperature distribution at device exit.

In 1994, Birk and Vandam published the results of scale model (1/4th) testing of DRES ball device, undertaken at the Exhaust System Test Facility (ESTF) for LM 2500 plume turbine prior sea trials. The scaling is based on Mach number similarity. Exhaust plume mass flow (1/16th of full scale) was generated using a hot air generator keeping the temperature of exhaust plume same as full scale. Though the scale model was fabricated from sheet metal, surface roughness was maintained same as full scale system. The study concluded that full scale device under sea conditions performed better than scale model under test conditions and theoretical predictions.

Thompson and Vaitekunas (1998) proposed a comprehensive and systematic approach to infrared suppression onboard modern ships by carrying out a detailed susceptibility analysis, trade-off studies and cost benefit analysis. In addition to passive suppression of plume and hot exhaust ducts, the paper explored use of special paints, water wash and water mist for reducing IR signature of solar heated hull surfaces.

Leon Galle and Schleijpen (1998) analysed naval ship survivability at sea from agile, intelligent and highly sensitive Anti-Ship Missiles (AShMs) in a four part series covering two critical aspects of ship survivability viz., susceptibility and vulnerability. Part II of the series provides an overview of factors contributing to IR signature onboard naval ships and highlights few suppression features installed onboard Royal Netherlands Navy Frigates.

In 2001, Munro, Van Dam and Birk compared three passive technologies viz., DRES Ball, Eductor-BLISS and Eductor-diffuser for plume turbines and 1000 KW diesel engines of Canada's DDH 280 destroyers. Eductor–diffuser design was observed to provide higher view angle protection, adequate plume cooling, lesser top weight and a cost effective option.

In 2003, Stephen W. Dudar developed a one-dimensional model of an eductor system with enhanced mixing which can be fitted within the space constraints of plume turbine exhaust of the Landing Helicopter Assault Ship Replacement (LHA (R)) class ships. The developed system provided 46% additional reduction in exhaust temperature compared to the DDG 51 class. Changing the aspect ratio of the mixing tube exit plane to 6:1 was observed to reduce infrared signature of plume by 20%.

Schleijpen and Neele (2004) carried out studies on the use of cold water spray for cooling of hot exhaust plume from plume turbines for the Royal Netherlands Navy and compared them with passive cooling using ambient air. The paper discusses spectral and integrated values of infrared emissions of cooled and uncooled plumes. Advantages of using sea water for exhaust plume cooling are presented.

Neele, Wilson and Youern (2005) observed that the standard method of specifying temperature reduction or contrast radiant intensity to be achieved often leads to problems due to weather conditions during sea trials being vastly different from those specified at design stage. Also, it is often difficult to implement actions to reduce signature as multiple agencies are involved. The new method proposes to define infrared signature suppression to be achieved in terms of constructional parameters such as emissivity of paints or thermal properties of insulation etc., which are quantitative, concise and verifiable.

Davis and Thompson (2006) presented an integrated approach to providing IR specification, be it a frigate or a tank or an aircraft. The approach recommended analysis of unsuppressed condition as baseline followed by cost benefit studies on increasing the level of IR suppression. The specification should be either in terms of level of signature reduction required (modality is flexible) or it is to be in terms of hardware required to achieve desired suppression. In both cases, modalities for product acceptance should be clearly stated.

Kok, Steven Loke Yew (2012) analysed combat survivability components of a surface ship and interaction between their designs that would affect overall survivability of the ship. Integrated signature management is recommended to reduce susceptibility of ship.

Yong-Jin Cho and Dae-Eun Ko (2017) used Taguchi experiment method to analyse the variables affecting diffuser design. Surface average temperature of diffuser duct and exhaust plume at outlet are variables directly related to intensity of infrared signature in naval ships. They concluded that changes in diameter of diffuser outlet affect plume temperature while the number of diffuser rings affects the metal surface temperature of diffuser.

Zhongwei Chen and Guoying Yao (2018) opined that naval ship stealth has seen a paradigm shift from being a single physical field in a single frequency such as radar cross section or infrared to a composite guidance system. Providing a list of such composite guidance anti-ship missiles, the susceptibility of surface ships for detection is analyzed. The feasibility of designing superstructure, exhaust stack and surface coatings with dual stealth capability is discussed.

Choudhary, Kumar and Murty (2022) discussed use of water mist along with ambient air to achieve enhanced suppression of plume IR signature. Methodology to assess reduction in missile lock-on-range achieved with different levels of infrared signature suppression is discussed.

Most notable designs of passive IRSS technology known to have been used onboard ships of world navies as studied in the literature review are shown in Fig. 2.2.

2.3 EJECTOR DESIGN FOR SHIP EXHAUST SYSTEMS

In 1949, Wilsted, Huddleston and Ellis studied the effect of plume turbine exhaust plume temperature (primary-jet) on several ejector configurations to examine the feasibility of using cold-ejector data for predicting hot-ejector performance. The study concluded that for short mixing lengths, weight flow ratio of primary to secondary jet varies directly as the square root of ratio of primary to secondary temperature. For long mixing tube lengths, increase in primary jet temperature reduces weight-flow of secondary fluid due to increase in static pressure.

Pucci (1954) studied a simple ejector system in which the primary jet, exhaust plume from shipboard plume turbines, is discharged into a cylindrical mixing tube having cross-sectional area larger than the jet. This induced second air flowing parallel to the jet axis. Both the air streams are mixed prior discharge to atmosphere. The system was developed for the Bureau of Ships, U.S. Navy to replace ventilation blowers in machinery spaces. It was observed that ejector performance is dependent upon mixing tube length, area ratio and flow rate ratio. The study concluded that for low area ratio ejectors, optimum performance is achieved when length of mixing tube is approximately seven times mixing tube diameter. The study was carried out with primary jet at ambient temperature due to test rig limitations.

As an extension of work of Pucci, Mitchel (1958) studied the design parameters for subsonic air to air ejectors with exhaust plume of plume turbines as the driving fluid. The dimensional parameters considered are specific to ventilation of ship machinery spaces. The study refined the one dimensional flow model by determining the momentum correction factor as a function of primary to secondary flow rate and temperature ratios. A static pressure correction factor was introduced to account for non-uniform static pressure at ejector exit. Initial study reported on straight. He concluded that an ejector-mixing tube combination gives best performance when standoff equals to 1.35 times nozzle exit diameter and mixing tube inlet to nozzle exit area ratio is 1.835.

Reddy and Kar (1968) estimated the ideal pumping efficiency of a water jet pump at unity flow ratio is 50%. He concluded that for maximum efficiency, jet pump with fixed nozzle should have an area ratio of 0.33 to 0.543 and should be operated with standoff distance of 1 to 2 nozzle exit diameter.

Vyas and Kar (1975) carried out experimental investigation of entrainment by an air jet ejector. The distances of jet exit from mixing tube was varied and it was observed that entrainment decreases as standoff increases from 2 to 6 times nozzle exit diameter. Air entrainment increased for mixing tube inlet to nozzle exit area ratios from 36 to 183. Mixing tube length of 44 times nozzle exit diameter is observed to provide complete mixing of primary and secondary flows. The study concluded that for Reynolds number (Re) greater than 10^4, entrainment ratio is independent of Reynolds number.

Decher (1981) studied aircraft turbofans with high aspect ratio nozzles and observed that infrared emissions can be reduced by a factor of 2 by keeping nozzle aspect ratio of 8. They observed that bypass ratios near unity reduce infrared emissions while after burning increases infrared emissions by 10 times compared to thrust. Increase in thermal efficiency reduces infrared emissions.

Chih-Ming Ho and Ephraim Gutmark (1987) observed that a small aspect ratio elliptic jet (2:1) entrains considerably higher than a circular jet. Secondary fluid is entrained mostly around the minor-axis plane due to moving away of the vortex core from jet axis. Flow properties such as turbulence intensity, mean velocity and Reynolds stress are observed to be considerably different in the minor and major axis planes.

Skebe, Mccormick and Presz (1988) carried out experimental work on circular and lobed ejectors with varying mixing tube length and reported about 200% entrainment with lobed nozzle. The study also reported that lobed nozzles enabled achieving mixing between primary and entrained fluids within half the MT length of circular ejectors.

Carletti and Rogers (1995) carried out experimental observations that large scale vortices introduced at nozzle exit of smaller and shorter circular ejectors increased the centreline velocity decay and enhanced the mass entrainment by 40%.

Tsan-Hsing Shih et al., (1995) proposed a new realizable k-ε eddy viscosity model and compared the predictions with both standard k-ε eddy viscosity model and available experimental data. The new model showed considerable improvement over standard model.

Hui Hu et al., (1999) studied the performance of seven combinations of rectangular lobed ejector mixing tube systems using three lobed rectangular nozzles and four rectangular mixing tubes. The experimental study was carried out on a low speed exhaust ejector system in lab. They reported improvement in pumping ability by 200-300% and reduction in mixing tube length by 33 to 50%. Rectangular aligned lobed nozzles had better pumping ability compared to rectangular staggered lobed nozzles albeit with higher pressure losses.

Mi, Nathan and Luxton (2000) carried out experimental study of nine differently shaped jets that include circular, elliptic, triangular, rectangular and cross shaped jets for turbulence characteristics and centreline mean flow. Non-circular jets demonstrated rapid decrease in centreline mean velocity indicating higher entrainment of ambient air. Isosceles triangle shaped jet with AR of 2.6 produced maximum mixing in the study. It is highlighted that jets of star, square and cross shape cause comparatively minimum changes in the far field mixing rates.

You-Hong Liu (2002) carried out numerical simulation and experimental measurement of secondary air entrainment by a circular lobed nozzle with and without central plug. Nozzle with central plug entrained 60 to 70% more secondary air and allowed the secondary air to reach core of primary jet resulting in better cooling and uniform thermal profile at exit of mixing tube. The improved performance of lobed nozzle with central plug comes at the cost of 0.8 to1.0% increase in back pressure.

Ghanshyam Singh, Sundararajan and Bhaskaran (2003) carried out experimental investigation of entrainment characteristics of a jet fluid from a nozzle with circular/ non-circular section entering a mixing tube having larger inlet area with and without standoff distance. The study observed enhancement in entrainment up to 30% with shift in jet location and validated the theoretical analysis of Pritchard et al. For smaller mixing tube diameters without standoff, circular jets entrained more air. The study developed a new correlation by modifying Pritchard's relation for confined circular jets to incorporate the effects of non-circular jets and standoff distance.

Wang and Li proposed new type of IRS device consisting of a lobed nozzle and double walled diffuser. The double-wall is envisaged to reduce radiation from walls compared to the single wall. Entrainment of secondary air and cooling of the gap between the walls was observed to improve when the diffuser is tilted from 0 to 10 degrees. The study reported a good match between experimental measurements and numerical simulation.

Václav Dvořák (2007) used a simple optimization algorithm for design of a lobed nozzle. Fluent 6.1 was used to study the effect of size and number of lobes for optimal secondary mass flow rate. The study observed that the objective function is not unimodal for a given shape of lobed nozzle. Too many lobes increased frictional losses and one of the local optimum is represented by the circular nozzle. Simulation studies indicated that each lobe generates two vortices with higher turbulent kinetic energy. This allows faster mixing but also causes a rapid rise in static pressure.

Maqsood and Birk (2007) carried out experimental and numerical study of round and oblong shaped ejectors with mixing tubes were bent at varying angles. Experimental tests were carried out for cold and hot conditions in a wind tunnel and parameters such as pumping ratio, pressure recovery and temperature ratio were studied. Numerical simulation was carried out to evaluate effectiveness of CFD software to predict the parameters. The study reported a decrease in pumping ratio and pressure rise with increase in bend angle. Ejectors with mixing tube at 45° angle recorded 28% less entrainment but performance improved with the addition of an entraining diffuser. Swirl in primary flow enhances entrainment upto 20° and decreases for higher swirl. Realizable k-ε turbulence model is recommended as more suitable to study air-air ejectors compared to Spalart-Almaras and k-ω models .

Yong Shan and Jing-Zhou Zhang (2009) carried out numerical simulation of different mixer configurations and observed that mixing efficiency is increased by over 65% using a lobed forced mixer resulting in 40% reduction in infrared radiation (normal to nozzle aspect) for a penalty of 3% reduction in thrust coefficient.

Mishra and Dash (2010) carried out a number of CFD simulations of exhaust plume flow through a cylindrical funnel with louvers and presented correlations to predict the suction rate of air flow into the funnel. The study concluded that the optimum diameter and height of funnel are 4.55 and 40 times nozzle dia respectively, for which mass suction into the funnel is highest. The study also established that the length of nozzle protrusion into the funnel has negligible effect on the entrainment of air into funnel.

Barik, Dash and Guha (2014) developed four different correlations to predict mass entrainment and two correlations to predict temperature at funnel exit by carrying out regression analysis using software program written in Engineering Equation solver. The data is obtained by carrying out numerous simulations by varying parameters such as nozzle exit Reynolds number, nozzle inlet diameter, diffuser ring overlap and standoff distance over range of operating conditions considered specific to the study. The correlations are validated with simulation results, experimental observations and similar results in literature.

Ju Hyun Im and Seung Jin Song (2015) presented a new analytical model based on control volume analysis and jet expansion model to characterize short ejectors that discharge non-uniform mixed flow at mixing tube exit due to incomplete mixing. Experimental tests were carried out with ejectors having length ratio of 2 for area ratio of 3.08 and length ratio of 3 for area ratio of 1.95. The study concluded that the proposed analytical model can accurately predict performance of short ejector. Velocity profile of entrained air at mixing tube inlet and exit predicted by the analytical model showed a good match with experimental measurements. Both simulation and experimental measurements of static pressure showed a kink when the outer shear layer reaches mixing duct wall. Secondary flow velocity was observed to decrease upstream of kink and increase downstream.

Sahu and Mishra (2015) carried out numerical simulation study to estimate the air entrained into a louvered horizontal cylindrical pipe with open and closed entrance. The entrainment of secondary air increased for both pipe configurations when area of louver is increased. The study reported a 1.1 times increase in amount of air entrained with six nozzles compared to single nozzle. Increase in pipe diameter increased air suction through louver for open entrance pipe while it decreased for closed entrance configuration.

L Singh, SN Singh and SS Sinha (2017) carried out systematic study using numerical simulation and concluded that optimum area ratio of nozzle exit to mixing tube inlet lies between 2 and 2.5. They observed that mixing tube length 8 times more than nozzle exit does not affect air entrainment. The study confirmed that Reynolds number (Re) above 10^5 does not affect entrainment.

Hamedi, Mahdavy and Jahromi (2018) carried out numerical simulation to study effects of convergence and divergence half-angles for compressible plume flows through a nozzle at different pressure ratios. The study reported a 6% decrease in discharge coefficient for increase in convergence angle from 5° to 40°. Further, a 3% decrease in nozzle gross thrust is observed for each 10° increase in convergence angle between 10°-40°. The study reported at convergence angle of 40°, flow separation occurs. The study report concluded that divergence half angle had no significant effect (1%) on the discharge coefficient.

Manoj and Sukanta (2019) carried out numerical simulation of exhaust flow through a thick hollow horizontal cylinder placed on ground and reported that the natural convective heat transfer from the inner surface of the cylinder increases initially with length to diameter ratio till a specific value and then it attains a constant value. The convective heat transfer from the outer surfaces increases continuously.

Mishra and Paramanik (2019) carried out experimental measurements of air entrainment into cylindrical mixing pipes of different diameters and length with variation in primary fluid jet location. Numerical simulations were carried out with both isothermal and hot jets. The study reported optimal nozzle protrusion of 1 cm for hot flow and 2 cm for isothermal jet. Highest entrainment was observed for cylinder of length 0.6 m and beyond this length the entrainment falls drastically. Entrainment increased with increase in pipe diameter.

Ganguly and Dash (2020) used a multi conical nozzle based IR suppressor with cylindrical nozzles arranged in a circle at the base. The study concluded that entrainment is optimum for five or six nozzles depending on plume temperature at nozzle inlet. Increase in overlap of funnels results in decrease in entrainment by 32% with consequent increase in plume temperatures by 40%.

Vikrant Chandrakar et al., (2022) carried out numerical simulation to study the natural cooling of exhaust duct housing infrared suppressor and concluded that time for cooling to ambient temperature is 39% lesser when both convection and radiation are considered together compared to considering only convection.

Dash and Barik (2023) proposed a new design of infrared suppressor by stacking conical and cylindrical funnels alternately. Flow pattern of entrained air and pressure variation are studied. Optimum diameter ratio of 1.1 is proposed for maximizing natural convective heat transfer from funnel walls.

Few designs of shipboard air ejectors as studied in the literature review are shown in Fig. 2.2.

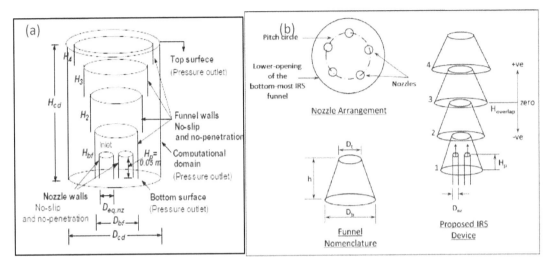

Fig 2.2 Cylindrical and Conical IRS Devices from Literature

(Source (a) Baril et al., 2014 (b) Ganguly et al., 2020)

2.4   ENTRAINING DIFFUSER DESIGN FOR SHIP EXHAUST SYSTEMS

Johnston (1954) carried out studies on annular diffusers having divergence angle from 6.5 to 15 degrees and an area ratio of 3.2. The study observed that increase in divergence angle resulted in non-uniform inlet conditions resulting in lower efficiency. Small angle diffusers with angles up to 6 degrees have up to 80% efficiency over a wide range of inlet velocity. At higher diffuser angles, the efficiency is reduced either by non-symmetry of flow or by low velocity region near walls.

Waitman, Reneau and Kline (1961) observed that the static pressure recovery coefficient of conical diffuser is affected by conditions such as velocity profile at inlet, turbulence intensity and swirl. The flow blockage parameter (ratio of displacement thickness to inlet width) at inlet correlated with diffuser inlet velocity profile.

Nicoll and Ramaprian (1970) reported significant improvement in conical diffuser performance even at moderate rates of annular secondary injection at inlet.

31

Analytical method used in the study to predict pressure recovery with inlet injection compared well with the experimental results.

McDonald, Fox and Dewoestine (1971) carried out experimental investigation on the effect of inlet swirl for wide angle conical diffusers (cone angle of 20∘). A total of 24 diffusers with divergence angles ranging from 4 to 31.2 degrees with area ratios 1.30 to 8.27 were tested for outlet flow profile. It was observed that inlet swirl had no impact on unstalled flows but has improved pressure recovery by 36% for badly stalled flows. The study recommended addition of swirl to improve overall performance of flow system for relevant cases.

Senoo et al. (1978) tested five conical diffusers with varying divergence angles to record the influence of swirl angle on pressure recovery. The study reported maximum pressure recover by a diffuser with 8 degree divergence angle for moderate swirl.

Klien (1981) studied large area diffusers and observed that pressure recovery is limited to 0.9 for thin boundary layer while higher values of turbulence improved pressure recovery.

Hoffmann and Gonzalez (1984) observed that a 2-dimensional diffuser with 20 degree cone angle showed an increase in pressure recovery of 20% for increase in turbulent intensity from 0.5% to 5%.

Singh, Agarwal, Sapre and Malhotra (1994) studied effect of inlet swirl for flow in equiangular annular diffusers. The study observed that inlet swirl enhanced pressure recovery, suppressed flow separation and reduced effective diffuser length.

Singhal et al., (2006) carried out numerical simulations on a non-conventional method to prevent separation of flow close to the wall of a wide-angle two-dimensional rectangular diffuser by injecting momentum through the moving surface using rotating cylinder(s) placed properly at the inlet of diffuser. The study reported that a single cylinder improved the pressure recovery by about 14% while two cylinders improved the recovery by 25%. Increased rotational speed of cylinders improved pressure recovery and flow distribution.

Rodrigo Queiroz Neiva et al., (2007) carried out experimental studies of water flow in two different diffuser geometries and captured the data using Particle Image Velocimetry (PIV). Numerical Simulation was carried out for the same conditions using software CFX. The study concluded that PIV is a good tool to study complex flows and that numerical simulation provides reasonably good match and also enables capturing additional data which cannot be measured experimentally.

Sen (2008) studied a two stepped conical diffuser having area ratio of 10 with divergence angles of 8° and 12° respectively. Based on experimental and computational studies, it was concluded that pressure recovery in the diffuser increases with slot angle, swirl and decrease in divergence angle. Mass entrainment through diffuser rings also increased with decrease in divergence angle, swirl and increase in interface height. The study concluded that the length of overlap does not affect pressure recovery but increase in wall thickness adversely affects pressure recovery.

Qi Chen and Birk (2009) carried out experimental study on pumping efficiency, temperature and velocity distribution of four-ring oblong entraining diffusers fed by oblong and straight air-air ejectors. The study was carried out with cold flow of 2.2 kg/s and hot flow of 1.8 Kg/s at 500°C. The study reported that oblong diffuser with oblong ejector has better mixing performance than with round ejector for low inlet swirl conditions. At 30 deg swirl, higher pumping efficiency is reported by oblong ejector while it degraded for round ejector. The study concluded that oblong ejector provided better overall performance for a short length ejector.

Sparrow et al. (2009) studied flow separation in diffusers with varying diffuser angles and observed that separation occurred at 7 deg expansion angle for laminar flows (Re< 2000). Increase in turbulence reduced length of flow separation.

Parminder and Sidh Nath (2014) carried out studies on square shaped stepped self-entraining diffuser for varying overlap conditions. The study concluded that slot overlap does not affect pressure recovery in the diffuser and that increasing the number of slots from 5 to 10 had no effect on cumulative mass entrainment. The study was for non-circular air ejectors with no standoff.

33

Ashok Barik et al., (2014) carried out numerical simulation studies of air entrainment into louvered funnel and reported a 70% increase in entrainment with a five time increase in louver opening area. The study recommended positioning the louvers at the bottom of funnel to achieve better entrainment. The study was carried out for lower Reynolds number flows ($3250 \leq \text{Re} \leq 6600$).

Yong- Jin Cho, Dae-Eun Ko (2017) examined four variables affecting diffuser performance, namely, diffuser outlet diameter, ring height, overlap and number of rings and optimized them into nine analysis cases using Taguchi experiment. Performance of IRSS was studied for plume and duct wall metal temperatures using CFD simulation for the optimized cases. The study concluded that exhaust plume temperature is greatly affected by diffuser outlet diameter while diffuser ring wall temperature is affected by the number of rings.

Lakhviner, Sidh Nath and Sinha (2017) proposed innovative guided slots for conical diffuser of an infrared suppression system and investigated its performance vis-à-vis non-guided slot diffuser for the same ejector. Guided slot diffuser entrained 3.5% more air and showed lower annulus temperature closer to the wall compared to non-guided slot diffuser. The study concluded that guided slot ejector-diffuser offers better suppression though pressure recovery is slightly lower.

Hardial Singh, and Arora (2019) reviewed the existing literature on the effect of diffuser geometry and inlet fluid conditions on output performance. The study concluded that optimum values of diffuser geometry (area ratio, divergence) and fluid conditions (Reynolds number, Mach number, turbulence intensities, swirl etc) with respect to pressure recovery are well established for 2-dimensional conical diffusers and certain annular diffusers.

Bhattacharya and Madhusudanachari (2019) proposed a variable diameter louvred cylinder in place of mixing tube and diffuser of standard IRSS device while keeping the total louver opening area same. The study examined changes in mass entrainment and pressure recovery for various parameters such as nozzle entry location, number of holes, hole location etc. It is concluded that optimum mass suction is obtained for a bottom open funnel with the nozzle positioned closer to the first row of holes.

Carrying forward the design work on the innovative diffuser with guided slots, Singh L et al., (2020) studied the effect of changing the angle of straight plate and observed an increase in cumulative mass entrainment from 2.88 kg/s at 0° to 4.04 kg/s at 28° with a corresponding reduction in wall temperature. This is attributed to the growth in size of cold pocket of ambient air below the straight plate with increase in angle. There is however, a corresponding fall in pressure recovery from 0.7 to 0.15 from 0° to 28°. The configuration of hybrid plate at first slot offered maximum entrainment.

Maheshwaran, Praveen and Vinoth (2020) carried out experimental measurements on two diffusers with divergence angles of 13° and 21° respectively. The study concluded that moderate swirl caused significant improvement in the stalled diffuser with 21° divergence but did not affect the diffuser with 13° divergence having incipient turbulent boundary layer separation.

James Ray Wright (2020) carried out a total of nine simulations on conical diffusers of varying diffuser half angle for swirl rates high enough to cause recirculation in the diffuser. The study concluded that for diffusers with recirculation, geometry has no effect on diffuser performance which becomes a function of swirl strength only.

Sachin et al., (2022) proposed a modified infrared suppressor with additional inward/ outward guides. The study observed that at optimal guide length of 0.163 nozzle dia and 15 deg inclination angle, air entrainment is maximum. Using nonlinear regression analysis, an empirical correlation equation is developed for air entrainment. Few designs of shipboard diffusers as studied in the literature review are shown in Fig. 2.3.

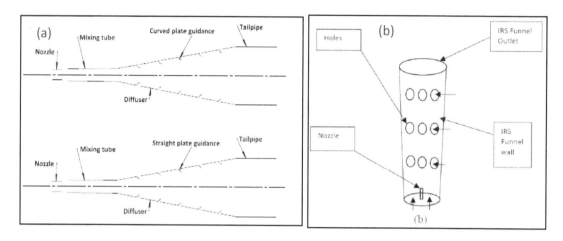

Fig 2.3 Entraining Diffusers from Literature

(**Source:** (a) Singh et al., 2020 (b) Bhattacharya and Achari, 2020)

2.5 STRUCRUAL AND FLUID PARAMETERS FOR ENGINE IRSS

Baham and McCallum (1977) discussed various techniques for design of exhaust funnel or ship stack that house the exhaust ducts from each engine, be it for propulsion or for power generation. Marine architectural issues related to superstructure design such as top weight, strength, safety, aesthetics etc., were considered along technical requirements related to smoke clearance of aft part of ship. Design guidelines for selecting the funnel height, shape, location of stack and techniques for estimating plume trajectory and downward wash are presented.

Fitzgerald (1986) described a step-by-step approach to predict exhaust plume dispersion from the stack for naval surface combatants. He presented the considerations for design of low-profile exhaust stack and predicted future direction in stack design.

Kulkarni, Namita and sheshadri (2005) investigated the effect that air wake from the superstructure has on exhaust plume dispersion from ship funnels. Wind tunnel studies were carried out to assess the effect of superstructure configuration for two velocity ratios. These studies funnel height for smoke clearance had an indirect impact on stack length available for infrared suppression systems.

Vijayakumar et al., (2008) carried out a holistic study on plume turbine exhaust flow from forward and aft funnels of a naval combatant with superstructure placed in between. The position of plume turbine intakes and other auxiliary equipment such as blowers was modelled. Hot exhaust plume at 500 °C was generated using heaters fitted in the funnel exhaust duct and mapping of thermal profile carried out with 1344 discrete measurements using RTDs. The study provided a vivid picture of infrared signature associated with the plume and secondary heating of superstructure due to plume impingement.

Vice Admiral GM Hiranandani (2009) traces the developments in the Indian Navy from 1991 to 2000 and discusses the contribution of research and development to self-reliance. The role of Under Water Ranges (UWR), Goa in measurement and management of ship signatures is discussed in detail.

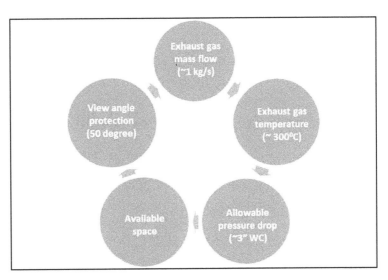

Fig 2.4 Shipboard Requirements for Ejector-Diffuser IRSS Systems

LM2500 technical data sheet (2017) of M/s GE Marine states that it is derived from aero-versions TF39 and CF6-6 and delivers 25060 KW at exhaust plume mass flow of 70.5 Kg/s and temperature of 566 °C. These plume turbines are used onboard frontline naval ships across the world for main propulsion. The plume turbines of Zorya Mashproekt, Ukraine have higher exhaust plume mass flow rates (~100 kg/s) at lower temperatures.

Technology Perspective and Capability Roadmap (2018) is a mission document that is updated periodically by the Government of India with a view to provide a broad perspective of the equipment, machinery and systems planned to be developed indigenously and inducted in the Army, Navy and Air force. The latest update promulgated (TPCR-2018) highlights the requirement of developing stealth systems including infrared signature suppression systems for naval ships.

Technical data sheet of MTU 20V4000 GS engine (accessed June 2023) installed onboard fast patrol boats of Indian Navy indicates exhaust plume temperatures of 460 °C at about 3 kg/s exhaust mass flow and power output of 2.5 MW. MoU signed between GRSE and MTU in Jan 2023 for indigenous manufacture of the MTU 4000 series engines indicates the validity of these parameters for all future naval projects.

Based on the data obtained from the engine suppliers of typical plume turbine and diesel engines used onboard naval ships world over, the range of fluid parameters and space constraints associated with ship exhaust systems are summarized as shown in Table 2.1.

Table 2.1 Typical Fluid and Structural Parameters of Ship Propulsion

Ser	Type of Propulsion	Primary Fluid	IRSS Length	IRSS Exit : Inlet Dia	Back Pressure
(i)	Main propulsion Boost (MPB)	80-100 kg/s @ 520 – 580 °C	8-10 m	1.8 -2.3	1245 Pa
(ii)	Main propulsion Cruise (MPC)	10-15 kg/s @ 450 – 520 °C	3.5 - 4.5 m	1.3-2	746 Pa
(iii)	Power Generation (PG)	1-6 kg/s @ 450 – 550 °C	2.5-3.5 m	2-4	746 Pa

2.6   TECHNOLOGY GAPS FROM LITERATURE

Based on literature review, following technology gaps have been identified:-

(a)   Non availability of tools to assess if optimal dimensional ratios recommended for air-air ejectors and diffusers in literature can meet space constraints of ships.

(b)   Methodology to predict aero-thermal performance of ejector-diffuser device that is employed for infrared signature suppression.

(c)   Limited studies on performance of conical subsonic air ejectors and entraining diffusers for infrared signature suppression viz., exhaust plume temperature at device exit, diffuser metal temperatures along its length and back pressure on engine.

(d)   Limited studies are carried out on experimental validation of numerical simulation results of infrared signature parameters.

2.7   SCOPE OF PRESENT RESEARCH

Based on the objectives of the research, literature review is carried out and technology gaps for ship infrared signature management are identified. Scope of the present research is as follows:-

(a)   Develop mathematical model for dimensional design of integrated ejector-diffuser that considers ship specific space constraints and exhaust flow conditions

(b)   Develop a tool for predicting the exhaust flow parameters within the developed dimensional design

(c)   Carryout computational study of different designs of nozzles, mixing tubes and diffusers to evaluate alternate dimensional designs and predict their performance.

(d)   Fabricate full scale IRSS devices and carry out experimental tests at Diesel Engine IRSS facility at NSTL.

2.8   CONCLUSION

Review of literature is carried out in four sections covering different aspects of the research area. Developments in the field of ship infrared signature suppression technologies since the 1930s are enumerated to highlight the fact information/ inputs required for design of infrared suppression systems are not adequately addressed in the literature, probably in view of their strategic nature. Optimal dimensional designs of subsonic air-air ejectors proposed in the literature are enumerated in second section and are reviewed as part of the present research to assess their suitability for shipboard applications. The third section deals with entraining diffusers considered to be suitable for installation in exhaust systems for pressure recovery and mass entrainment. The fourth section examines the fluid and structural parameters associated with typical propulsion systems used onboard naval ships. Technology gaps limiting design and development of practical systems for full ship infrared signature management are identified. Scope of present research is defined.

CHAPTER 3

MATERIAL AND TOOLS

3.1 INTRODUCTION

The chapter presents the materials and tools used in the research work. Based on technology gaps, it was considered essential to develop mathematical tools that enable Ship Builders to carry out preliminary dimensional design integrated ejector–diffuser technology based Infrared suppression device for class of ship specific parameters. Methodology adopted in development of two mathematical modeling tools is presented. Given the complex nature of fluid flows inside exhaust stack, numerical simulation is a powerful tool to gain deeper insights into aero-thermal performance of different designs based on chosen technology. Computational fluid dynamics tools are cost effective and require leaner resources in terms of time and materials. A brief description of the software tools used for numerical simulation is presented. Experimental measurements are required to evaluate and validate effectiveness of the software tools. For the research work, the experimental studies are carried out using test facilities available at Naval Science and Technology Laboratory (NSTL), Visakhapatnam. Full scale models of nozzles, mixing tubes and diffusers are fabricated and different combinations assembled on a specially designed test structure. Sensors and instruments are required to measure the performance. Brief description of the materials and tools used for carrying out the research are presented.

3.2 MATHEMATICAL MODELLING

Mathematical modelling is the process of defining a real life problem in mathematical terms to understand interconnected relationships, carrying out mathematical analyses, obtaining results and reinterpreting the model. It helps in scientific understanding of the process through quantitative expression of the knowledge available about the system. Sensitivity analysis is a part of the mathematical modelling process where in the effect of change in system output is studied with changes to one or more parameters. Mathematical modelling helps us to predict the likely performance of

41

a system prior to physically building and testing it. Mathematical modelling process used in the current study is based on the 'System Model'. In this method, the system is built by integrating a series of sub-models, each of which describes a real life process and the sub-models have interaction with each other. Two mathematical models, namely, Ejector-Diffuser for Ship Applications (EDSA)' for dimensional design and 'Aero-Thermal Performance Model (ATPM)' for predicting flow parameters are developed.

3.2.1 Building Mathematical Models EDSA and ATPM

Building a mathematical model necessitates having a clear understanding of the objectives for developing it. The various steps involved in building the model are as follows:-

(i) Real life systems are complex and are affected by a number of parameters. To reduce complexity of modelling, assumptions are made based on available knowledge about the system and the physics associated with it. While generating component level dimensional designs, the EDSA model considers the properties of plume to be same as that of hot air given the marginal effect of gas composition on temperature. Similarly, the ATPM model calculates bulk mean flow parameters at the outlet of each sub component viz., nozzle, mixing tube and diffuser rings.

(ii) Flow diagrams are used to describe the subsystems and the integration sequence. In the current study, there are 3 subsystems viz., nozzle subsystem, mixing subsystem and diffuser subsystem.

(iii)The subsystems are described in terms of mathematical equations. Some of the equations are formulated based on the principles of physics while others are based on empirical relations from literature review. Details of the same are presented in Chapter 4.

3.2.2 Sensitivity Analysis

One of the advantages of mathematical modelling is its ability to carry out sensitive analysis. It provides an understanding of the effect that variation in one parameter has on the upstream and downstream component designs and thus, establishes limiting dimensions of various components forming the system. In the current study, the dimensional parameters of the subsystems are varied and associated changes in fluid parameters within the IRSS are studied. This helps in identifying weakness in the modelling process especially with respect to areas using empirical formulae. It also helps in identifying critical dimensional parameters and the accuracy required in their design. While carrying out sensitivity analysis, caution is required with respect to correlated parameters because when changing one parameter, the correlated parameters also need to be changed.

3.3   NUMERICAL SIMULATION

Mathematical modelling provides alternate dimensional designs based on simplification of the process of exhaust flow within IRSS while solving the equations analytically. For example, analytical estimate of the average plume temperature at mixing tube or diffuser exit assumes complete mixing of the hot exhaust plume with ambient air. The plume profile, is therefore, uniform. However, in real life, the short lengths associated with ship based IRSS subsystems coupled with velocity of the fluid flow preclude uniform mixing. Such complex processes are best captured through software based simulation. Two most common applications of numerical simulation are Finite Element Analysis (FEA) and Computational Fluid Dynamics (CFD). While FEA is used to simulate real world physical structures in terms of geometry and load conditions, CFD is used to study flow dynamics by solving fluid flow equations. The current study focuses on use of CFD for simulation of exhaust flow within IRSS to predict parameters affecting the infrared signature of the ship. The software CATIA for generation of 3D models and Simcenter STAR CCM+ for CFD are used in the current study. Flowchart depicting the processes involved in numerical simulation using CFD is shown in Fig. 3.1.

3.4   MODELLING

The software used for generating 3D models for the current study is CATIA V 5. The abbreviation 'CATIA' stands for Computer Aided Three dimensional Interactive Application. It is a multi-platform software suite developed by the French company M/s Dassault Systems for the 'Mirage' fighter jet program and can be used for a number of engineering applications such as Computer Aided Design (CAD), Computer Aided Manufacturing (CAM), Computer Aided Engineering (CAE), 3D modeling and Product

43

Lifecycle Management (PLM). For 3 D modelling, the software has two options viz., solid and surface. In the current study, CATIA V 5 has been used. The flow chart depicting modelling process is shown in Fig 3.2.

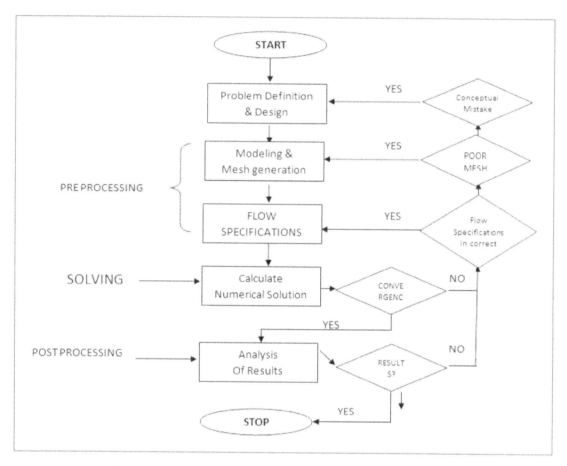

Fig 3.1 Workflow in Numerical Simulation using CFD

3.5   COMPUTATIONAL FLUID DYNAMICS

Fluid dynamics was studied predominantly through experimentation in the 17th Century and through theoretical methods in the 18th and 19th century. For most part of 20th century, fluid dynamics studies were either on pure theoretical or pure experimental basis. The advent of computers combined with development of accurate numerical algorithms for solving physical problems on them changed the paradigm and forms the genesis of a new approach called computational fluid dynamics. The first work using computational power to model fluid flow is considered to have been performed in 1967 at the Los Almos National lab, USA. Since then, the use of CFD has risen exponentially and today forms the basis for analyzing all complex fluid dynamics phenomena. It is important to understand that output of CFD simulations needs to be validated either with experimental observations or with previously performed analytical or empirical analysis of the problem.

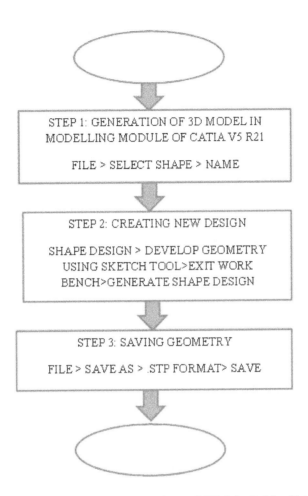

Fig 3.2 Flowchart depicting Generation of 3D Model in CATIA V5 R21

For the current study, CFD software from M/s Siemens, Simcenter STAR CCM+, is chosen because of its availability and its proven use in industry. The abbreviation 'STAR CCM+' stands for 'Simulation of Turbulence in Arbitrary Region Computational Continuum Mechanics.' The '+' is used to indicate that the program is developed in C++. The software provides a single integrated environment that includes CAD for 3 D modelling, automated meshing, choice of physics models specific to each problem and good post processing features that facilitate design exploration. However, a brought out at Para 3.8, the 3D modelling has been carried out using the software CATIA V 5 in view of its comparative ease of handling.

45

3.5.1 Creating Domain

Computational domain refers to the space in which the solution of a CFD simulation is estimated. The domain is discretized into a mesh or a grid to solve the discretized equations associated with fluid flows. The computational domain for internal and external flows varies due the area of interest. In internal flows, the geometrical shape forms the boundary and the space inside forms the computational grid. In external flows that simulate flow around geometry, domain is selected based on the area where flow solution is required. The domain shape and volume depend on the aerodynamics of the geometry under study. The procedure for creating the domain in STAR CCM+ is as follows:-

- Go to parts > new shape part > select a cylinder

- Specify the length and diameter of the cylinder

- Create a domain by selecting "create". The domain created is shown in Fig 3.3

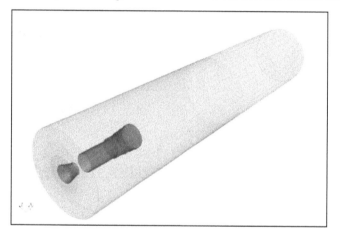

Fig 3.3 Creation of a domain around IRSS System

3.5.2 Meshing

The flow of fluids and associated heat transfer are governed by partial differential equations that are difficult to solve analytically, except for extremely simple cases. Therefore, to analyze the flow of fluids and heat transfer, the computational domain is divided into a number of small cells or elements that form a grid or a mesh as shown in Figure 3.6. The governing equations of the fluid flow are discretized and are solved inside each cell and the results are passed onto the adjacent cells. The discretization procedure approximates the space and time derivatives of the flow variable at each node to algebraic functions. It is usually based on finite volumes, finite elements or finite difference methods. To get a complete view of fluid flow across length and breadth of the domain, care must be taken to ensure proper continuity of the solution as it moves across common interface surfaces between the cells. The size and shape of the cells within the computational domain need not be uniform and are defined based on the region of interest. The quality of mesh is extremely important in CFD as it has significant effect on the results. It is defined in terms of three parameters, namely cell aspect ratio, skewness and smoothness. Aspect ratio is the ratio of longest to the shortest side in a cell. For best results, it should be equal to 1. Larger aspect ratios result in interpolation errors of unacceptable magnitude. Also, the size of adjacent cells should not vary by more than 20% to minimize local variations and ensure smoothness. The factors that determine the quality of mesh for a given problem include required precision of results, rate of convergence, computational time and grid independence requirements.

3.5.3   Meshing Models

In STAR CCM+, meshing can be done in two ways viz., Parts based or Region based. In the simulation tree, Parts based meshing is available under Geometry > Parts > Operations > Automated Mesh. In the simulation tree, Regions based meshing based is available under Continua>models> various meshers are available. Each Region covers some area of the geometry, which may encompass only section of a Part or may include more than one Part. To assign parts to regions, one is to right-click on a part upon which a selection menu appears and helps to assign parts and their surface to regions. The 'Regions' also define inlets, outlets and walls of the simulation. Prior to mesh generation, inlet and outlet boundary conditions need to be assigned correctly so that volume mesh propagates successfully. Each part is to be assigned with one surface mesher and one volume mesher. The various meshing models available in Star CCM+ are shown in Fig. 3.4. With all the meshers, it is important to check both the size of the cells and the target size to ensure the mesh will be fine enough to capture most of the geometry.

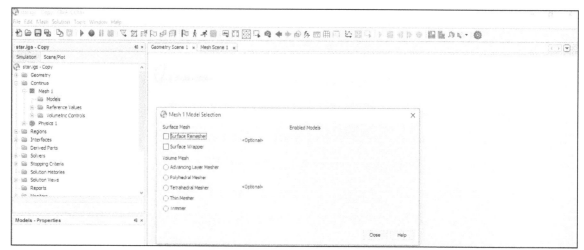

Fig 3.4 Mesh Models in STAR CCM+

There are two types of Surface Mesh available.

- **Surface Wrapper:** For complex geometry, the surface wrapper covers the initial surface and ensures that the geometry is closed and is of sufficient quality for generating surface and volume meshes.

- **Surface Re-mesh:** This option Re-meshes the initial surface to provide a quality discretized mesh that is suitable for CFD. It is used to re-triangulate the surface and also provides support to omit specific surfaces or boundaries for preserving the original triangulation from the imported mesh.

The types of volume mesh are as follows:-

- **Trimmer:** This type of volume mesh is recommended when an underlying custom mesh needs to be used or if the surface quality is not good enough for a polyhedral mesh. It is also useful in modelling external aerodynamic flows due to its ability to refine cells in regions of boundary separation such as wake. It generates a volume mesh by cutting a hexahedral template mesh with the geometry surface

- **Tetrahedral mesh:** It generates a volume mesh composed of tetrahedral-shaped cells. These meshes are recommended only when comparison is to be made with legacy tetrahedral models.

- **Polyhedral mesh:** The volume mesh is generated using polyhedral-shaped cells. This mesh is more accurate, less diffusive and more stable than an equivalent tetrahedral mesh. For a given surface, it contains 5 times fewer cells compared to a tetrahedral mesh.

- **Advancing layer mesh:** The mesh creates a surface mesh on the wall and projects it to create prismatic cell layers next to wall boundaries and a polyhedral mesh elsewhere. The prismatic cell layers help capture the boundary layer, turbulence effects, and heat transfer near wall boundaries.

- **Thin mesh:** This mesh is used where good quality cells are required to capture the solid material thickness adequately. It generates a prismatic layered volume mesh for thin geometries.

The other mesh models are available as optional and are selected on a case to case basis for specific purposes as enumerated.

- **Prism layer mesh:** Add prismatic cell layers next to wall boundaries.
- **Extruder:** It is typically used for inlet and outlet boundaries to extend the core volume mesh beyond the original dimensions of the starting surface, so that a more representative computational domain is obtained by extruding.

- **Generalized cylinder mesh:** For elongated cylindrical regions, it generates appropriate volume mesh. It uses extruded prismatic cells to reduce the overall cell count and improve the rate of convergence in some cases.

- **Shelling mesh:** The mesh is specifically used for modelling casting methods. It generates a shell mesh region from a boundary that one of the core volume mesh has meshed.

- **Embedded thin mesh:** Similar to the default thin mesh, it is also used to generate a prismatic type mesh in predominantly thin geometries.

The window for selection of optional meshed models along with core mesh is shown in Fig. 3.5. The selected ones are called the enables models.

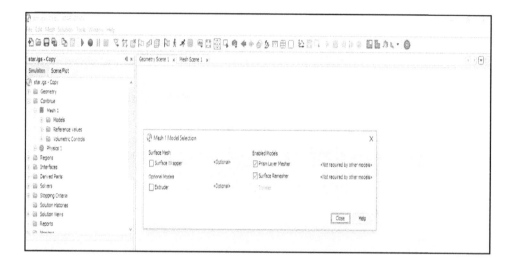

Fig 3.5 Selection of Optional Mesh Models with Core Mesh

3.5.4 Volumetric Controls

To reduce the computational time while preserving the accuracy of results to a reasonable level, the computational domain has mesh that extends from very fine meshing to coarse meshing, depending on the precision levels of the data that needs to be captured.

3.5.5 Reference Values of Meshing and Grid Independence

Values of all the parameters related to meshing can be specified on selecting the option 'Reference values.' These include Base Size, Automatic surface repair, CAD projection, Maximum cell size, Surface curvature, Surface growth rate, Surface Proximity, Surface size and Template growth rate. The quality of grid affects the solution. A high quality grid provides accurate solutions but requires large computational time and resources. Low quality or poorly resolved grid may provide solutions faster but can give incorrect solutions. Grid independence study is a process by which the Base Size of the mesh is increased in steps say, by 20% and the solution is compared to the previous result. When the variation in value of critical parameters of the solution is not significant (which may vary for each case), the solution is considered to be grid independent. Good grid not only ensures a reliable solution but may also facilitate early convergence.

3.5.6 Meshing Process

The flowchart depicting the process used in STAR CCM+ for generation of meshing in the computational domain is shown in Fig 3.6.

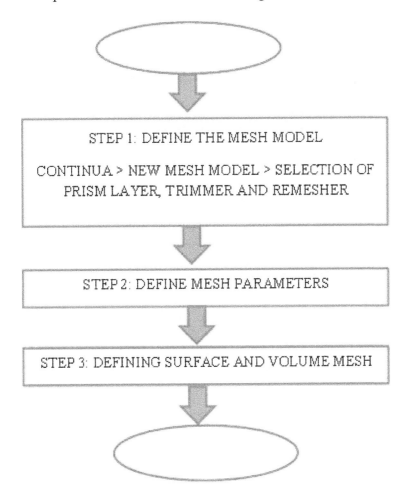

Fig 3.6 Flowchart of Mesh Generation in Simcenter STAR CCM+

3.5.7 Defining Physics Models

The fluid parameters of the flow such as the inlet conditions such as pressure, temperature, velocity etc., and the mathematical formulation defining the flow are selected as Physics models. In STAR-CCM+, the physics models are defined on a physics continuum. *Continua > Physics 1 > Select Models.* The process of selecting the physics models is shown in Fig 3.7. For the current study, the medium or material is Plume and hence, the same is selected.

In STAR CCM+, option is available to activate 'auto-select recommended models' by checking the relevant box. The 'Physics Model Selection' dialogue box guides in selecting the type of flow and in assigning properties to it. The physics models selected for the current study and their relevance is discussed as part of numerical simulation studies in Chapter 5. The reference values associated with parameters selected in the physics continuum is defined under *Continua>Physics 1>Reference Values*. These include maximum allowable temperature, maximum allowable wall distance, reference pressure etc. The surroundings of the IRSS system are considered as initial conditions for the simulation and are defined under *Continua>Physics 1>Initial Conditions*. Typically, the initial conditions pertain to pressure, static temperature, velocity components and turbulence parameters. The simulation is saved. On completion of the selection process, the color of the Physics 1 node turns from grey to blue.

Fig 3.7 Procedure to Select Physics Models

3.5.8   Defining Boundary Conditions

The boundary conditions are specified under *Regions>Boundary Conditions*. For the current study, some of the boundary conditions include pressure boundary in the domain, velocity boundary at the IRSS system inlet, wall of ducts as baffle.

3.5.9   Governing Equations

Computation fluid dynamics mainly depends upon three fundamental governing equations of fluid dynamics

- Mass conservation   (Continuity equation)
- Momentum Conservation (Navier-Stokes Equation)
- Energy Conservation   (Energy equation)

3.5.10 Run the Simulation

The above process completes the pre-processing process as indicated in the flowchart at Fig 3.5. The numerical simulation can now be run under *Initialize Solution>Run Solution*.

3.5.11 Post Processing

Post processing is carried out to study the effect of flow parameters such as temperature, mean flow velocity, pressure distribution, film cooling etc. The results are presented in terms of thermal, velocity and pressure profiles at desired cross sectional planes.

3.6   EXPERIMENTAL SET UP

3.6.1   Hot Air Generator Facility

The experimental set up consists of a Hot Air Generator (HAG), exhaust duct, fuel tank, control panel as shown in Fig 3.8 to generate primary fluid and motive force. Relevant specifications of the test facility are shown in Table 3.1.

Table 3.1 Specifications of Hot Air Generator

Ser	Parameter	Description
(a)	Type	Diesel Oil Fired Gas Generator
(b)	Number of Passes	02
(c)	Hot air mass flow	$0.5 - 2.5$ m^3/s
(d)	Temperature Range	150 - 420 °C
(e)	Inlet Duct diameter	0.318 mm
(f)	Maximum MT diameter	425 mm
(g)	Height of IRSS	0.8 m $- 2.1$ m
(h)	Flame sensor	UV type detector

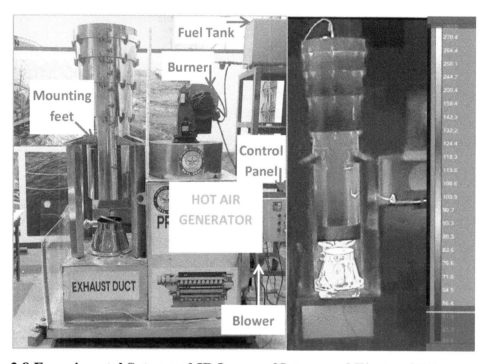

Fig 3.8 Experimental Setup and IR Image of Integrated Ejector-Diffuser IRSS

3.6.2   Gas / Hot Air Generator

The gas generator is a device that produces exhaust gases similar to that of propulsion/power generation engine onboard naval ships. It draws ambient air into the combustion chamber using a fan into which fuel is injected. The gas generator used for the current study produced exhaust gas mass flow ranging from 0.5 Kg/s to 1.5 Kg/s at temperatures ranging from 100 to 300 °C by adjusting the amount of fuel injected and motor speed.

3.6.3 Ejector-Entraining Diffuser IRSS Device

IRSS system considered for study comprises of a nozzle, mixing tube and diffuser. The nozzle is mounted on the exhaust duct of the gas generator by suitable flange fitting. The mixing tube and the diffuser are mounted on a supporting structure using mounting feet. The gap between nozzle and mixing tube is referred to as standoff distance. The nozzle, mixing tube and standoff together comprise the ejector subsystem while the multitude of concentric rings comprises the entraining diffuser. Onboard ships the nozzle, mixing tube and diffuser sections are attached to the funnel walls through tie rods with suitable rod end bearings that can absorb vibrations and torsional stresses.

3.6.4 Principle of Operation

The primary gas stream (exhaust gas) from the engine is accelerated through the nozzle as shown in Fig 3.9. Ambient air is naturally sucked into the exhaust gas flow through 'stand-off' due to negative pressure and cools the hot plume by mixing in the mixing tube. The combined flow is decelerated in the diffuser to recover kinetic energy. However, negative pressure persists even in the diffuser section and ambient air is sucked through ring gaps which not only creates an air film over diffuser ring surfaces but also cools exhaust gas. The pressure at device exit is near ambient.

3.6.5 Structural and Fluid Parameters for Study

The choice of fluid parameters at nozzle inlet and permissible structural dimensions are based on the capacity of experimental facility available at Naval Science and Technology Laboratory (NSTL), Visakhapatnam. The structural and fluid parameters considered for mathematical modelling, numerical simulation and experimental studies for the research work are shown in Table 3.2.

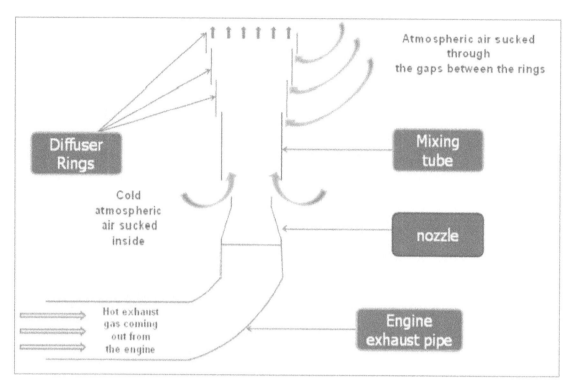

Fig 3.9 Principle of Operation of Ejector-Entraining diffuser IRSS Device

Table 3.2 Input Structural and Fluid Parameters for Study

Ser	Parameter	Symbol	Value
(i)	Exhaust plume mass flow at nozzle inlet	$\dot{m}_{ni}$	1 kg/s
(ii)	Temperature of primary fluid	T_{ni}	300 °C
(iii)	Ambient air temperature	T_{amb}	30 °C
(iv)	Primary fluid pressure at inlet	P_{ni}	1.014 bar
(v)	Pressure loss within IRSS	P_b	0.430 bar
(vi)	Nozzle inlet diameter	$D_{ni,}$	318 mm
(vii)	Total length of IRSS	L	1664 mm
(viii)	Diffuser exit diameter	D_{de}	478 mm
(ix)	View Angle Protection	θ	$\geq$ 50 deg

3.6.6 Experimental Study Methodology

The methodology followed for experimental study is as follows:-

(a)   Flange mount nozzle of select shape on existing exhaust duct of the hot air generator.

(b)   Mount the mixing tube and diffuser sections on the mounting feet of support structure as shown in Fig 3.8.

(c)   Check concentricity with plumb line to ensure centerline is maintained with exhaust ducting.

(d)   Mount instrumentation viz., RTDs on diffuser section, thermocouples on diffuser exit and pitot tube at ejector inlet

(e)   HAG is to be gradually loaded and operated at 1 kg/s exhaust mass flow and temperature of 300 °C.

(f)   Carry out measurement of duct wall temperatures (02 nos. RTDs @ each location), Exhaust gas temperature (05 Thermocouples) at diffuser exit and back pressure at minimum 05 locations across nozzle inlet plane.

(g)   Carry out infrared imaging using Thermal imaging camera

3.7   FABRICATED MODELS

Fabrication of select IRSS models is undertaken to study the effects of component level design changes on performance of integrated ejector-diffuser IRSS. The fabricated models are round, cylindrical and 4 lobed nozzles as shown in Fig 3.10. Innovative mixing tube that can be used in single mixing tube (SMT) and twin mixing tube (TMT) configuration as seen in Fig. 3.11. Two different diffuser designs, one having 4 rings and the other having 2 rings are seen in Fig 3.12. The 2 ring diffuser has provision to adjust the rings for 3 positions, namely, no overlap, one ring gap overlap and two ring overlap.

Fig 3.10 Nozzles of different shapes and Inlet-Exit Area Ratios

Fig 3.11 Single and Twin Mixing Tube Configurations

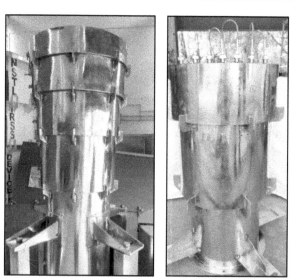

Fig 3.12 Four Ring and Two Ring Diffuser Configurations

3.8   INSTRUMENTATION AND SENSORS

Temperature sensors and the pressure measuring devices are fitted at designated locations as shown in Fig 3.13 to record the parameters related to IR signature. Static pressure at nozzle inlet is measured using L-type pitot tube. Technical specifications of the sensors and instrumentation used in the current study are enumerated in succeeding paragraphs.

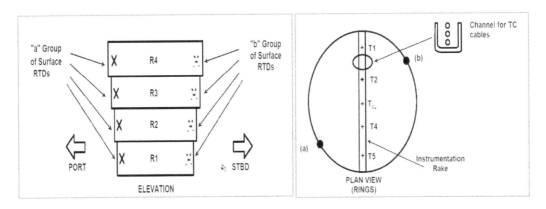

Fig 3.13 Location of RTDs on Diffuser Ring Walls and TCs at IRSS Exit

3.8.1   Resistance Temperature Detector (RTD)

RTDs are mounted at mid-point of each diffuser ring on the outer side to measure duct wall temperatures. Design of typical RTD used in the study is shown in Fig 3.14 and specifications are as follows:-.

- Type   : Washer Type RTD (PT-100)
- Range   : 0-200°C
- Measuring Parameter : funnel surface temperature
- Cable length   : 25 meters
- Output measurement : 12 channel indicator
- Accuracy   : 100.00 ±0.12 O at 0°C
- Standard   : DIN/IEC 60751 (or IEC751) Class B

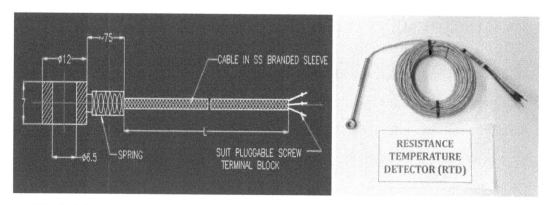

Fig 3.14 Washer Type PT 100 RTD for Exhaust Diffuser Ring Temperatures

3.8.2 Thermocouple Specification

Thermocouples are mounted at diffuser ring exit for measurement of exhaust plume temperatures. Design of typical thermocouple used in the study is shown in Fig 3.15. Specifications are as follows:-

- Type   : K-TYPE
- Range   : 0-600 ^{O}C
- Standard   : ANSI/ASTM E230 or IEC 60584 (Class 2)
- Measuring Parameter   : Plume or gas temperature
- Measuring Location   : Top of the funnel
- Cable length   : 25 meters
- Output measurement   : Temperature transmitter/converter
- Accuracy   : $\pm$ 1^O C

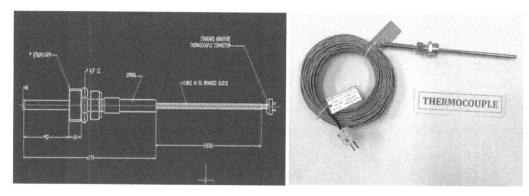

Fig 3.15 K-Type Thermocouple for Exhaust Plume Temperature Measurement

3.8.3   Pressure Sensor

Static and dynamic pressures associated with exhaust flow through an ejector-entraining diffuser IRSS are measured using an L Type pitot tube. Typical pitot tube used in the study is shown in Fig 3.16. Specifications are as follows:-

- TYPE   : S-TYPE/ L- Type
- Range   : 0 to 2 PSI (0-55 Inches of water column)
- Measuring Parameter : static pressure
- Output measurement : Pressure transmitter/converter (model: C9500)
- Accuracy   : $\pm$ 1" H_2O

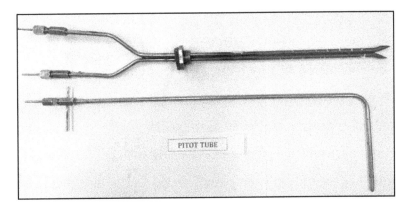

PITOT TUBE

<u>Fig 3.16 S and L-Type Pitot Tubes for Static and Dynamic Pressure Measurements</u>

3.8.4   Infrared Imaging Camera

To capture the thermal profile of the exhaust ducts, it is not possible to instrument the full length of the diffuser rings. An infrared imaging camera helps in capturing the profile which can then be analyzed with the software provided by the supplier. Thermal Imaging systems work by detecting the heat energy being radiated by the objects and require absolutely no light. Infrared Thermography is the technique used for producing a visible image of the infrared light emitted by objects due to their thermal condition. The image produced by an Infrared Camera is called a 'Thermogram' or sometimes a 'Thermograph. IR camera captures radiosity of the target being viewed. Radiosity is defined as the infrared energy coming from a target modulated by the intervening atmosphere and consists of emitted, reflected and sometimes transmitted IR energy. An opaque target has a transmittance of zero. The colours on an IR image vary due to variations in radiosity. The radiosity of an opaque target can vary due to the target temperature, target emissivity and reflected radiant energy variations. Objects with high emissivity imply most of the radiosity is due to the target and hence, for objects with emissivity below 0.5, radiosity measurements are not considered accurate. For opaque objects, if the emissivity and the background (reflected) temperature are known, an IR camera with a temperature measurement feature (radiometric camera) can give temperatures accurate to within a few percentages. To get the temperature value, the IR camera extracts just fraction of the radiosity due to energy emitted by the target. The reflected component is subtracted and the result is scaled by the target emissivity. The resulting value is then compared to a calibration table and temperature is extracted.

Emissivity is generally considered a material surface property. The exhaust duct surfaces are considered as gray body. Grey body has an emissivity less than one that is constant over wavelength. A real body has an emissivity that varies with wavelength. The calibration tables assume the material to be Grey body, which is true in most cases. In addition, it is also affected by the shape of the object, viewing angle, wavelength and temperature. In case of semi-transparent materials, the thickness also affects emissivity. The exhaust plume, on the other hand, is a spectral radiator and is seen only with the use of special filters. IR imaging camera of make FLIR model T 1020 camera as shown in Fig.3.17 is used in the study and technical specifications are placed in Table 3.3.

Table 3.3 Specifications of Thermal Imaging Camera used in Study

Make and Model	FLIR T 1020
IR resolution	1024 × 768 pixels or more
Thermal sensitivity/NETD	<20 mK @ +30°C (+86°F) or better
Field of view (FOV)	28° × 21° or less
Focal Length	Suitable focal length to give minimum FOV
Lens identification	Automatic
Image frequency	30 Hz or more
Focus	Manual & automatic

Detector type	Focal plane array (FPA), un-cooled micro bolometer
Spectral range	LWIR range (8-14 μ)
Temperature range	0 to 600 °C
Temperature range setting	Can be adjusted in several temperature range based on requirement
Setting	Emissivity and other parameters like reflected temperature, distance from the target etc., can be adjusted in the camera.

Fig 3.17 Thermal Imaging Camera FLIR T 1020 used in Study

3.9   CONCLUSION

An overview of various materials, methods and tools used in carrying out analytical, computational and experimental work related to the research work are presented. Methodology adopted in development of two mathematical modeling tools used for preliminary design of IRSS systems is presented. Work flow of numerical simulation tool STAR CCM+ which is used in the research to gain deeper insights into the complex nature of fluid flows in ship exhaust system are discussed. Details of experimental facility available at Naval Science and Technology Laboratory, Visakhapatnam, materials used in the current study such as hot air generator, scale models, infrared imaging cameras, sensors and instrumentation is provided.

CHAPTER 4

MATHEMATICAL MODELING AND DIMENSIONAL DESIGN

4.1 INTRODUCTION

The design of conventional exhaust ducts for marine propulsion and power generation systems is an established art. The introduction of infrared signature suppression systems has brought in complexity in estimating the flow parameters at various stages of the ejector-diffuser system. Mathematical models available in literature are not adequate to completely characterize the flow through an integrated ejector-diffuser installed on ship exhaust system. With signature management assuming greater significance over the years, target infrared signature suppression to be achieved has become part of the Build specifications. Need has therefore arisen to have a tool that will ascertain the overall dimensions of IRSS device required to meet specified IR suppression targets so that space requirements can be considered while designing the funnels. The chapter focuses on development of mathematical tools that enable dimensional design of IRSS devices for ship specific parameters and calculate aero-thermal parameters associated with flow through an integrated ejector-diffuser device.

4.2 TYPICAL ARRANGEMENT OF EXHAUST SYSTEM ON SHIPS

Onboard naval ships, the main propulsion systems are located in engine rooms at the bottom part of the ship. Exhaust ducts carry hot exhaust plume from the engine to funnel exit. Typical arrangement of exhaust system from engine outlet to funnel exit including the infrared suppressor in respect of gas turbine and diesel ships is shown in Figures 4.1. The exhaust ducts have thermal insulation lagging on the outer side for safety of those operating close vicinity. Hence, flow of hot exhaust gas can be considered to be adiabatic with minimal convective and radiated heat losses. Methodology for estimation of exhaust flow parameters in conventional exhaust ducts

64

such as pressure loss through bends, head loss, friction loss etc., is an established practice. Infrared suppressors are typically installed at the top end of exhaust stack and are flush with the exit of funnel as shown in Fig 4.1. Typically, each exhaust duct has its own suppressor and hence, the number of suppressors is dependent on the number of exhaust ducts housed inside.

The introduction of infrared suppressor has complicated the design of ship exhaust systems in two ways. Structurally, they have larger openings compared to exhaust ducts provided by engine supplier and hence, funnel dimensions cannot be estimated without preliminary design of IRSS devices. These devices induce large amounts of ambient air into the hot exhaust flow path making available tools inaccurate for estimation of exhaust flow parameters at funnel exit.

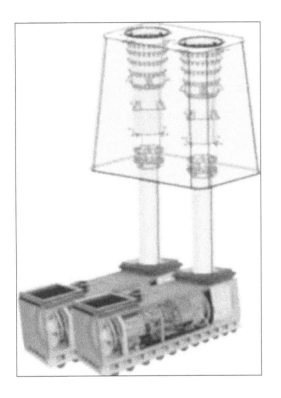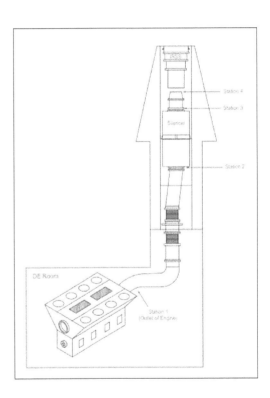

Fig. 4.1 Exhaust Duct Arrangement onboard (a) Gas Turbine & (b) Diesel engine

4.3   INPUT PARAMETERS FOR MATHEMATICAL MODEL

The input parameters required for design of ship specific integrated ejector-diffuser infrared suppressor are exhaust flow (fluid) parameters (temperature, mass flow and back pressure allowance by engine supplier) and funnel space constraints (inlet duct diameter, stack length available for IRSS, maximum permissible exit diameter and view angle protection required). Factors affecting component design/ optimization for shipboard applications are discussed in the chapter.

4.4   DESIGN OF EJECTOR SUBSYSTEM

The ejector system consists of nozzle, standoff distance and mixing tube.

4.4.1   Nozzle

The design the nozzle is based on the permissible pressure loss in the system. For convergent nozzles with gas flows at sub sonic speeds (<0.3M), density of the flow does not vary. The rapid flow in the Ejector can be approximated as adiabatic and favourable pressure gradient makes the flow one dimensional with very little frictional losses. Onboard marine platforms, the nozzle is typically flange mounted on top of the exhaust duct and hence, its inlet diameter is governed by the diameter of the exhaust duct provided by engine supplier. For Ejector-Diffuser type of IRSS with large stand-off distance, the pressure loss is entirely due to dynamic pressure of the nozzle. Therefore, nozzle exit diameter is designed to maximize exit velocity and obtain optimum pumping capacity. The nozzle exit diameter and length can be arrived at using equations 4.1 to 4.3.

$$\left(\frac{dW}{dt}\right)_{system} = 0 \qquad (4.1)$$

$$\frac{dp}{\rho} + vdv + gdz = 0 \qquad (4.2)$$

$$Nozzle\ length = \frac{Nozzle\ inlet\ dia - nozzle\ exit\ dia}{\tan \alpha/2} \qquad (4.3)$$

4.4.2   Standoff Distance

The standoff distance (S) is the space provided for passive induction of large amounts of secondary/ ambient air due to pressure difference and affects the cooling of the hot gas in the mixing tube. Based on literature review of liquid/liquid jet pumps, it is observed that for exhaust flows with large turbulence (high Reynolds number), the standoff distance is a function of nozzle exit. It is seen from literature review that different values of optimal standoff distance are reported by different studies such as 1.35 D_{ne} (Mitchel, 1958), as 0.8 $D_{ne}<S<2D_{ne}$ (Mueller, 1964; Reddy and Kar, 1968 and Bonnigton and King, 1972), as $2D_{ne}<S<3D_{ne}$ by (Sen, 2008) and as 2.25D_{ne} (Singh et al., 2018).

For the current study, the standoff is considered from the equation

$$0.8\ D_{ne} < S < 2\ D_{ne} \qquad (4.4)$$

4.4.3 Mixing Tube

The mixing of hot exhaust gas with secondary air induced through standoff distance is facilitated in the mixing tube. Though longer mixing tubes facilitate better mixing, studies in literature recommended keeping mixing tube length less than three times the diameter. It is observed that the optimum area ratio of mixing tube to nozzle exit is recommended in literature as 1.835 (Mueller, 1964), as 2.25 (Singh, 2018) and as 1.76 (Ashok et al., 2014). Considering funnel space constraints, equations 4.5 and 4.6 are recommended for generating alternate dimensions of mixing tube

$$2 < L_{mt}/D_{mt} < 3 \qquad (4.5)$$
$$2 < A_{mt}/A_n < 3 \qquad (4.6)$$

4.5   DESIGN OF DIFFUSER SUBSYSTEM

The diffuser subsystem consists of a number of concentric rings that are connected each other with brackets. In the literature, diffuser design studies focused on effect of inlet swirl on pressure recovery and mass entrainment as also optimizing diffuser half angle. Optimal diffuser divergence half angles proposed are 6.5 to 15 degrees (Johnston, 1954),   8 degrees (Senoo et al., 1978) and 15 deg (Sachin et al., 2022). For ship applications, the length of diffuser is optimized for view angle and diffuser ring duct metal temperatures.

4.5.1 Overall Diffuser Dimensions

View angle is defined as the angle (θ) subtended by the horizontal at mixing tube exit with the top of diffuser at diametrically opposite end as shown in Fig 4.2. It defines the level of IR protection i.e. the acceptable range of detection by missile IR seeker. The diffuser length is designed to keep view angle equal to or greater than defined value, thus, keeping the duct metal surface temperatures lower. Further, the diffuser length is also limited by maximum diffuser half angle (β) to avoid flow separation.

The study recommends design of diffuser based on equations 4.7, 4.8 and 4.9.

$$L_d = Tan\ \theta\ \frac{(D_{de} + D_{mt})}{2} \tag{4.7}$$

$$L_d = \frac{D_{de} - D_{mt}}{2\ Tan\ \emptyset} \tag{4.8}$$

$$L_d = \frac{D_{de}\ (Tan\ \theta)}{(1 + tan\theta tan\beta)} \tag{4.9}$$

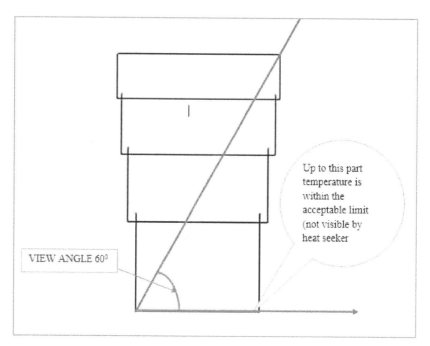

Up to this part temperature is within the acceptable limit (not visible by heat seeker

VIEW ANGLE 60°

Fig 4.2 View Angle Protection for Infrared Suppression

4.5.2 Diffuser Ring Length and Slot Width

The length of each ring is designed to be such that entrained air forms a continuous unbroken film on its inner surface. The gap between rings (slot gap) and ring length can be determined from empirical relations (Mukherjee, 1976) for estimating film cooling efficiency of tangential air injection assuming a fully developed turbulent boundary layer, ideal slot geometry and no pressure gradient . The actual lengths would then be fine-tuned through numerical simulation.

4.6   EJECTOR - DIFFUSER SHIP APPLICATION (EDSA) MODEL

The research proposes to develop a mathematical model to design an ejector-diffuser IRSS for ship applications (EDSA) that takes ship specific fluid parameters (temperature, mass flow and back pressure allowance by engine supplier) as input and applies funnel space constraints (inlet duct diameter, stack length available for IRSS, maximum permissible exit diameter and view angle protection required). The model generates alternate dimensional combinations of nozzle, mixing tube and diffuser as output.

4.6.1   Input Parameters for EDSA

Considering the dimensional and fluid parameters available at the experimental test facility, input parameters considered for the mathematical model are shown in Table 4.1. For shipboard applications, these parameters are available in the technical data sheets as indicated in literature review for LM2500 gas turbine of M/s GE, USA and 4000 series diesel engine of MTU, Germany. Other required input parameters are also provided by the engine supplier. The space constraints are provided by the shipbuilder. For the same funnel, height available for diesel engine IRSS is comparatively shorter than gas turbine because of the need to accommodate silencer to meet noise requirements onboard

Table 4.1 Input Parameters for EDSA

PARAMETER	VALUE/ RANGE OF VALUES		UNIT
INPUT PARAMETERS			
Mass flow rate of exhaust gas (ṁ)	1.00		Kg/s

Temperature of exhasut gas (T₁)	300		°C
Dia of exhaust duct at inlet (Dni)	0.318	Typical values associated with	m
Dia of exhaust duct at funnel exit (Dde)	0.478	power generation equipment onboard	m
Maximum permissible IRSS length (L)	1.664	ships/ Experimental test set up at NSTL	m
Nozzle Convergence Angle (α)	15		deg
Back pressure limitation (P₁-P₂)	746		Pa
Envisaged no. of rings (N)	4		Nos.

4.6.2   Calculated Parameters

For the set of input parameters specified in Table 4.1, EDSA calculates flow parameters required for dimensional design of the nozzle, standoff, mixing tube and diffuser. Some of the parameters are shown in Table 4.2. It may be noted that the pressure loss considered for design of nozzle (720 Pa) is slightly less than allowable limit (746 Pa) to keep a margin of safety.

Table 4.2 Structural and Fluid Parameters for EDSA

CALCULATED PARAMETER	VALUE/ RANGE	UNIT
Area at nozzle inlet (A_1)	0.079	m^2
Density of exhaust gas (ρ_1)	0.616	Kg/m^3
Velocity of exhaust gas at nozzle inlet (V_1)	20.445	m/s
Mach Number at nozzle inlet	0.040	m/s
Max. Pressure loss in nozzle (P_1-P_2)	726	Pa
Losses due to nozzle convergence	0.036	Pa
Dynamic viscosity (μ)	0.00002917	Kg/m.s
Velocity of exhaust gas at nozzle exit (V_2)	52.67	m/s
Reynolds Number (Re)	137322.60	m/s

4.6.3 Optimal Component Dimensional Range

The range of output dimensional parameters permissible for each of the components and the choice of optimal range considered is indicated in Table 4.3. The combination of values to be chosen for integrated device is an iterative process that involves evaluating the aero-thermal performance of each design

Table 4.3 Range of Output Dimension parameters provided by EDSA

Dimensional Parameter	Symbol	Flow of Dimensional Design	Min Dimension	Chosen Dimension	Max Dimension	Unit
Exit diameter of nozzle	D_{ne}	$(2(P_1-P_2)/\rho+V_1^2)^{0.5}$	0.318	0.220	0.217	m
Length of nozzle	L_n	$(D_{ni}-D_{ne})*Tan(90-\alpha)/2$	0.182	0.220	0.277	m
Standoff distance	S	$(0.8\,D_{ne}<S<2\,D_{ne})$	0.176	0.220	0.441	m
Mixing tube diameter	D_{mt}	$(2\,A_{ne}<D_{mt}<3\,A_{ne})$	0.312	0.336	0.382	m
Mixing tube length	L_{mt}	$(2D_{mt}<L_{mt}<3D_{mt})$	0.623	0.662	1.145	m
View Angle	θ	$(>50°)$	50	54.3	60	deg
Diffuser length (View Angle criteria)	L_d	$Tan\,\theta*(D_{de}+D_{mt})/2$	0.471	0.566	0.744	m
Diffuser length (Total length criteria)	L_d	$(D_{de}+D_{mt})/2Tan\beta$	0.682	0.561	-0.198	m
Equitable ring length	L_r	L_d/N	0.171	0.140	-0.050	m
Equitable ring gap	R_g	$(D_{de}-D_{mt})/N$	0.017	0.014	0.008	m

4.6.4 Study of Nozzle Design using EDSA

Studies have been carried out to understand the contribution of three parameters, namely, convergence angle, nozzle length and nozzle exit area on the back pressure that is likely to be imposed on the engine. Static pressure at nozzle inlet due to each of the parameters is calculated and is shown in Table 4.4. Effect of variation of nozzle inlet to exit area on back pressure and convergence angle observed for current study is plotted in Fig. 4.3

Table 4.4 Nozzle Design Study for EDSA

Mass Flow (ṁ)	Nozzle Inlet Dia	Nozzle Length	Nozzle Convergence Angle (°)	Gas Vel at Inlet (V₁)	Gas Vel at Exit (V₂)	$V_1^2-V_2^2$	Back Pressure P_2-P_1	Area ratio	Nozzle Exit Dia
1.00	0.32	0.22	1.00	20.43	20.68	10.23	4	1.01	0.316
1.00	0.32	0.22	4.00	20.43	21.45	42.88	15	1.05	0.310
1.00	0.32	0.22	7.00	20.43	22.27	78.78	26	1.09	0.305
1.00	0.32	0.22	10.00	20.43	23.14	118.37	38	1.13	0.299
1.00	0.32	0.22	13.00	20.43	24.07	162.21	51	1.18	0.293
1.00	0.32	0.22	16.00	20.43	25.06	210.94	66	1.23	0.287
1.00	0.32	0.22	19.00	20.43	26.13	265.31	83	1.28	0.281
1.00	0.32	0.22	22.00	20.43	27.27	326.26	102	1.33	0.275
1.00	0.32	0.22	25.00	20.43	28.50	394.89	123	1.40	0.269
1.00	0.32	0.22	28.00	20.43	29.83	472.58	147	1.46	0.263
1.00	0.32	0.22	31.00	20.43	31.28	561.01	174	1.53	0.257
1.00	0.32	0.22	34.00	20.43	32.86	662.24	205	1.61	0.251
1.00	0.32	0.22	37.00	20.43	34.59	778.89	241	1.69	0.244
1.00	0.32	0.22	40.00	20.43	36.49	914.24	283	1.79	0.238
1.00	0.32	0.22	43.00	20.43	38.60	1072.48	332	1.89	0.231
1.00	0.32	0.22	46.00	20.43	40.94	1259.04	389	2.00	0.225
1.00	0.32	0.22	49.00	20.43	43.57	1481.01	458	2.13	0.218
1.00	0.32	0.22	52.00	20.43	46.53	1747.77	540	2.28	0.211
1.00	0.32	0.22	55.00	20.43	49.89	2071.97	640	2.44	0.203
1.00	0.32	0.22	58.00	20.43	53.74	2470.93	763	2.63	0.196
1.00	0.32	0.22	61.00	20.43	58.19	2968.80	916	2.85	0.188
1.00	0.32	0.22	64.00	20.43	63.38	3599.96	1110	3.10	0.181

| *Inlet Area* | *0.0795* | *Density* | *0.616* | *(Z2-Z1) *ρg* | *1.33* | | | | |

It is seen that for the conditions associated with the study, back pressure is likely to increase significantly for area ratios beyond 2. The study also concluded that the effect of convergence angle and length of nozzle with respect to back pressure imposed on engine is considerably lower than nozzle exit area.

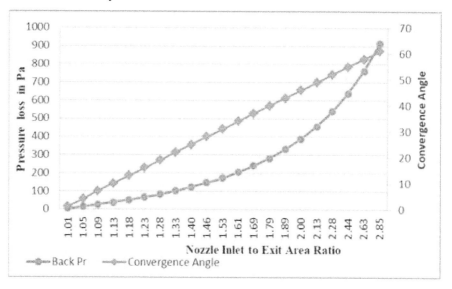

Fig 4.3 Effect of Nozzle Inlet to Exit Ratio on Back Pressure

4.6.5   Study of Diffuser Design with EDSA

The effect of view angle on diffuser design is studied. It is seen that diffuser length (L_d), exit diameter (D_{de}) and mixing tube diameter (D_{mt}) are a function of the view angle protection (θ). The effect of increasing view angle protection on the ratio of diffuser length to exit diameter and diffuser length to mixing tube diameter are shown in Fig. 4.4. It is observed that for view angle protection beyond 65 degrees, there is a steep increase in diffuser length to diffuser exit and mixing tube exit diameter ratios. Thus, for ejector-diffuser type IRSS systems, view angle protection above 65 degrees will mandate additional funnel height.

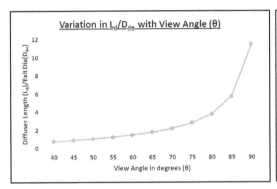 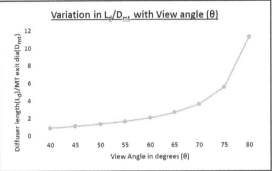

Fig 4.4 Effect of View Angle on Diffuser & Mixing Tube Design

4.7   AERO-THERMAL PERFORMANCE MODEL (ATPM)

For a given set of fluid and structural parameters at inlet to nozzle, a number of alternate components dimensional designs can be generated which will meet the space constraint requirements of funnel. However, there is need to assess them for meeting the infrared signature suppression targets. Towards this, the current research has developed a mathematical model to predict aero-thermal performance (ATPM) of integrated ejector- diffuser IRSS. ATPM takes DDR generated component dimensions and fluid parameters at nozzle inlet as input and calculates performance parameters of integrated ejector-diffuser IRSS viz., plume temperature along IRSS, fluid velocities at inlet & exit of each component, secondary and tertiary air induced etc.

4.7.1 Input parameters for ATPM

Dimensional design of an ejector-diffuser IRSS obtained using EDSA mathematical model considered as input parameters for the ATPM is shown in Fig 4.5

Table 4.5 Input Parameters of ATPM

Ser	Parameter	Value	Unit
1	Mass Flow of Exhaust at Nozzle Inlet	1.000	Kg/s
2	Temperature at Nozzle Inlet	300	°C
3	Temperature of Ambient Air	30	°C
4	Diameter at Nozzle Inlet	0.318	m
5	Diameter at Nozzle Exit	0.220	m
6	Diameter of mixing tube	0.318	m
7	Diameter of Ring 1	0.384	m
8	Diameter of Ring 2	0.416	m
9	Diameter of Ring 3	0.450	m
10	Diameter of Ring 4	0.478	m

4.7.2 Utility of ATPM in Preliminary Evaluation of EDSA Generated Designs

To understand the ATPM capabilities, the performance of a given combination of mixing tube-standoff-four rings diffuser is studied for the choice of a round nozzle in terms of three inlet to exit area ratios viz., RN1 (A_{ni}/A_{ne} = 1.1), RN 2 (A_{ni}/A_{ne} = 2.1) & RN 3 (A_{ni}/A_{ne} = 3.1). The flow parameters along the IRSS system for each of the combinations as calculated by the ATPM model is shown in table 4.5.

Prima facie, it is seen that RN 3 provides considerably lower temperatures than RN 1. However, ATPM also provides values of back pressure on engine viz., 1125.69 Pa for RN 3, 433 Pa for RN 2 and 5 Pa for RN 3. It is now observed that the back pressure for IRSS with RN 3 is beyond permissible value and will affect engine performance adversely. Thus, ATPM allows elimination of dimensional combinations that are unlikely to meet shipboard constraints. The mathematical model ATPM is only for preliminary screening.

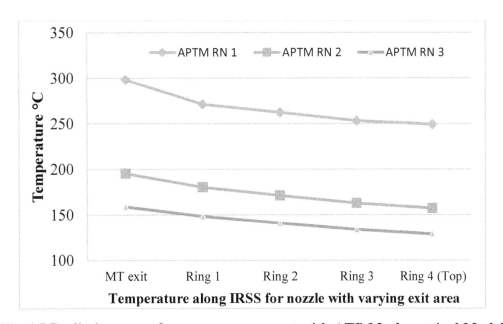

Fig 4.5 Preliminary performance assessment with ATP Mathematical Model

Limitations associated with ATPM include assumption of complete mixing of fluids within available component length, inability to predict exhaust duct thermal profile, non-consideration of nozzle exit shape etc., mandate use of numerical simulation for evaluating alternate integrated IRSS designs for optimal IR signature reduction.

Table 4.6 Evaluation of Nozzle Exit Area Design with ATPM

Ser	Parameter	Unit	ATPM RN 1	ATPM RN 2	ATPM RN 3
1	**Mass Flow of Exhaust at Nozzle Inlet ($\dot{m}_{in}$)**	Kg/s	**1**	**1**	**1**
2	**Temperature at Nozzle Inlet**	°C	300	300	300
3	Temperature of Ambient Air	°C	30	30	30
4	Diameter at Nozzle Inlet	m	0.318	0.318	0.318
5	Diameter at Nozzle Exit	m	0.315	0.220	0.180
6	Diameter of mixing tube	m	0.336	0.336	0.336
7	Mass flow of Secondary Air at MT inlet	Kg/s	0.007	0.634	1.096
8	Mass Flow of Primary fluid at MT exit	Kg/s	1.007	1.634	2.096
9	Temperature at MT exit	°C	298	195	159
10	Diameter of Ring 1	m	0.384	0.384	0.384
11	Mass flow of tertiary air at Ring 1	kg/s	0.112	0.164	0.184
12	Mass flow of primary fluid at Ring 1 exit	Kg/s	1.119	1.798	2.280
13	Temperature at Ring 1 exit	°C	271	180	148
14	Diameter of Ring 2	m	0.416	0.416	0.416
15	Mass flow of tertiary Air at Ring 2 inlet	Kg/s	0.044	0.115	0.159
16	Mass flow of primary fluid at Ring 2 exit	Kg/s	1.163	1.913	2.438
17	Temperature at Ring 2 exit	°C	262	171	141
18	Diameter of Ring 3	m	0.45	0.450	0.45
19	Mass flow of tertiary Air at Ring 3 inlet	Kg/s	0.047	0.122	0.168
20	Mass flow of primary fluid at Ring 3 exit	Kg/s	1.210	2.035	2.606
21	Temperature at Ring 3 exit	°C	253	163	407
22	Diameter of Ring 4	m	0.478	0.478	0.478
23	Mass flow of tertiary Air at Ring 4 inlet	Kg/s	0.021	0.086	0.124
24	Mass flow of primary fluid at Ring 4 exit	Kg/s	1.231	2.121	2.730
25	Total Air induced into IRSS System ($\dot{m}_{suc}$)	Kg/s	0.231	1.121	1.730
26	**Mass Flow Average Temp at IRSS exit**	°C	249	157	129

4.8 INFERENCES FROM STUDY USING EDSA AND ATPM

The mathematical tools EDSA and ATPM provided range of component dimensions permissible for ship applications while meeting infrared suppression requirements. The effect of changes in sub component level critical parameter and its effect on performance of integrated system is summarized in Table 4.6.

Table 4.7 Critical Parameters of Ejector-Diffuser IRSS Device

Ser	Subsystem Parameter	Significance	Performance Indicator
1.	Nozzle exit diameter	Affects system pumping capacity and device length	Back Pressure and velocity at nozzle exit
2.	Standoff distance	Pumping efficiency of nozzle; affects cool air available for mixing	Induction of Secondary air through Standoff
3.	Mixing tube diameter	To ensure smooth entry of secondary flow while catering for primary fluid divergance	Induction of Secondary air through Standoff
4.	Mixing tube length	To enable mixing between primary and secondary fluids	Bulk mean temperature at Mixing tube exit
5.	Ring height difference	To ensure uniform pressure gradient along the length of IRSS device	Pressure gradient along length of device
6.	Number of rings	Affects the production cost machining requirements and diffuser exit diameter	Diffuser exit diameter
7.	Ring gap	Volume of secondary flow for effective film cooling	Total secondary air in Diffuser
8.	Overlap length	Affects the film formation and the number of rings required	Surface average ring temperature and peak temperature

4.9 COMPETING DESIGN CONSIDERATIONS FOR INTEGRATED IRSS

Shipboard application demands trade-off due to contradictory requirements between funnel space constraints and IRSS system performance, few of which are highlighted below.

(a)   Smaller exit diameter of nozzle increases primary fluid velocity and higher pressure drop resulting in larger amount of secondary and tertiary air induction. However, it also increases back pressure on the engine.

(b)   Larger standoff distance allows more ambient air for cooling the hot exhaust plume but also affects the length of mixing tube and diffuser sections.

(c)   Longer mixing tube length enhances mixing and lowers the plume temperature. But for a given stack length, this reduces length of diffuser and thereby, view angle protection.

(d)   Longer diffuser ring length reduces number of rings and consequently, diffuser exit diameter, but results in higher metal temperatures as film cooling effectiveness decreases.

(e)   Larger ring gap allows more ambient air but increases the diffuser exit diameter.

4.10 UTILITY OF MATHEMATICAL TOOLS

Design of ships has traditionally used the telescopic model that follows a design spiral comprising of four phases, namely, concept design, preliminary design, contract design and detail design. Ship builder is required to undertake a series of activities that repeat for each phase with increasing level of detailing. Thus, at the conceptual stage, there is a requirement to formulate overall dimensions of the funnel. With signature management assuming greater significance over the years, target infrared signature suppression to be achieved has become part of the Build specifications. Need has therefore arisen to have a tool that will ascertain the overall dimensions of IRSS device required to meet specified IR suppression targets so that space requirements can be considered while designing the funnels. It is observed from literature review

4.11   CONCLUSION

The naval fleet consists of a number of ships that include aircraft carriers, cruisers, destroyers, frigates, corvettes etc., Based on the overall length, width and tonnage of the ship, main propulsion and power generation systems of varying capacity are fitted onboard. Ship Builders need to estimate overall dimensions of the funnel as part of Concept Design. Presently, mathematical tools that provide a quick data of overall dimensions of infrared signature suppressor that can meet targeted suppression levels are not available. The research work envisages developing a generic tool that meets this critical requirement.

CHAPTER 5

NUMERICAL SIMULATION STUDIES

5.1   INTRODUCTION

In this Chapter, the effect of component level design changes on the performance of integrated ejector-diffuser IRSS is studied using numerical simulation. The component design changes are studied through simulation of 12 case studies that include three nozzle exit diameters and six nozzle exit shapes, two designs of mixing tubes, four and two ring diffusers. The component dimensions for each combination of ejector and diffuser are obtained from EDSA mathematical model and are used to generate three dimensional surface models using CATIA V5 R21. Simulations are carried out using Computational Fluid Dynamics (CFD) solver STAR CCM+. The output parameters of the simulations are specific to infrared signature of ships and thus, augment the available literature on ejectors and diffusers.

5.2   GENERATION OF 3D MODELS

Based on the dimensional designs generated through mathematical modelling, 3D surface models for all the case studies are developed in CATIA V5 R21 as per procedure brought out at Para 3.4. 3D models of all the 12 case studies are in Figures 5.1, 5.2, 5.3 and 5.4 respectively. The computational domain is created in CATIA and then imported to the CFD software Simcenter STAR CCM+ as '.STP' file.

5.3   COMPUTATIONAL DOMAIN

The computational grid of typical integrated ejector-diffuser comprising lobed nozzle, standoff, mixing tube and four rings diffuser and the far field is shown in Fig 5.1. The dimensions of the computational domain are arrived at through an iterative process by carrying out a Domain Independence study. For the current, study of an ejector-diffuser device, it is observed that width of 4 times the largest diameter (diffuser exit) and length of four times the height of IRSS is sufficient to ensure that disturbances at the boundaries do not affect performance of integrated ejector-diffuser Infrared suppressor.

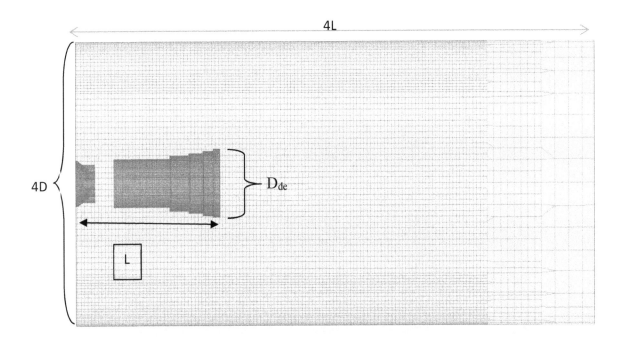

Fig 5.1 Computational Domain of Passive IRSS System

5.4   GRID STRUCTURE

Major portion of the computational domain is built in CATIA and is imported into STAR CCM+ where rest of the geometry is constructed. Meshing of the model is undertaken in the CFD software. Surfaces are created at top section of each of the rings and mixing tube to help in better visualization of images during post processing. Five types of grid volume shapes viz., trimmer (hexahedron), tetrahedral, advancing layer mesher, thin mesher and polyhedral are available in Simcenter STAR CCM+ for 3D meshing. It is observed that the structured mesh made up of hexahedral elements having lower skewness gives better numerical solution convergence and has less numerical diffusion than unstructured tetrahedral mesh. Progressive mesh is generated for the whole geometry including domain.

The domain is divided into four volume shapes as shown in Fig 5.2. The first volume shape covers the geometry of IRSS system and has very fine mesh generated in that volume shape. The second shape covers the volume around the IRSS system. Since the effect of physical changes play a major role in this area too, fine mesh is generated in this volume shape. The third volume shape covers areas farther to the IRSS system with limited effect of physical changes. As a compromise between larger domain and lower computational time, the mesh in this volume shape is unstructured mesh. For IRSS systems, it is extremely important to capture the physics near the walls accurately. Hence, Prism layer is generated near the walls.

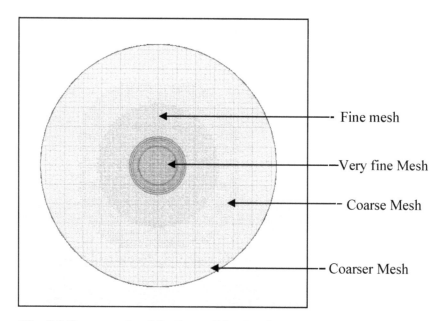

- Fine mesh
- Very fine Mesh
- Coarse Mesh
- Coarser Mesh

Fig 5.2 Progressive Mesh used in the Study

To capture the physics accurately, mesh with hexahedral elements is used. Tetrahedral cells are used in the corners given their adaptability to take shape. Five different base diameters viz., 500 mm, 450 mm, 400 mm, 318 mm and 280 mm are studied for variation in mass flow average exhaust plume temperature at diffuser exit. Details of the mesh and grid independence study for case study 2 are in Table 5.1.

Table 5.1 Details of Mesh for Computational Domain of IRSS System

Base Size →	500 mm	450 mm	400 mm	320 mm	280 mm
Plume Temp. at Diffuser Exit	151.6 °C	151.8 °C	157.5 °C	153 °C	157.2 °C
Ambient air induced	1.214 kg/s	1.211 kg/s	1.113 kg/s	1.189 kg/s	1.118 kg/s
No. of Cells	8,55,366	11,28,703	15,68,981	29,67,822	42,48,360

Initially, a base size of 500 mm was assumed and the solution converged quickly. The base size was then decreased in steps resulting in higher number of cells in the domain and hence, higher computational time. When the grid was coarse, the mass flow average temperature fluctuated considerably (3.62%) from its preceding value. As the grid became finer, the solution took longer but variation from preceding value became smaller (0.13%). It is seen that increase of Base Size from 320 mm to 280 mm resulted in 43% increase in number of cells and associated run time. Thus, it can be seen from Grid Independence study that a Base Size of 320 mm achieves the best compromise between reliable solution and higher resource consumption.

The mesh at 320 mm Base Size consisted of 29,67,822 cells, of which 28,81,051 (97%) are hexahedral cells, 1220 are tetrahedral cells (0.04%) at corners, 4 prism layers and a 10 degree skew angle. Meshing parameters for other case studies have comparable Base Size with minor variation in number of each type of cells.

5.5   CFD SOLVER

Simcenter STAR CCM+ version 10.02.012 is used as solver and post processor.

5.5.1   Physics models

Physics model in STAR-CCM+ defines how a physical phenomenon in a continuum is represented. They define the primary variables of the simulation (such as pressure, temperature, and velocity) and the mathematical formulations used to generate the solution. An appropriate combination of models is necessary for the complete definition of the physics continuum. Flow is defined as three dimensional steady flow of plume. Since the actual composition of exhaust plume is dynamic in nature depending on the fuel used, it is defined as an ideal plume. The velocity of the flow at inlet to IRSS system is approximately 20 m/s which is much lower than 0.3 M. ATPM model calculated a maximum velocity of 55 m/s in the mixing tube which is also considerably lower than 0.3 M. Thus, the flow is considered as incompressible and segregated flow model is selected for solving. Physics model selection window of STAR CCM+ software is shown in Fig 5.3. The physics models selected for the current study are shown in Fig 5.4.

5.5.2 Turbulence Modelling Approach

First step in choosing the turbulence model is to calculate the Reynolds number. The Reynolds number associated with Case 2 baseline IRSS is 136973 and that of all other cases is closer to this. It can be concluded that the flow is turbulent in nature.

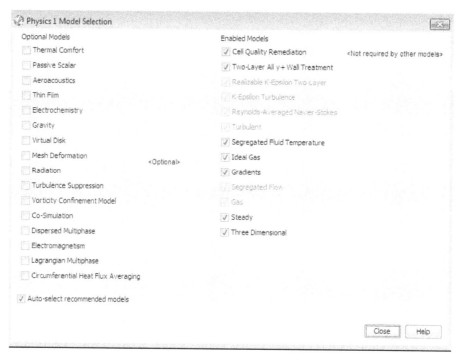

Fig 5.3 Physics Models Selection window in Simcenter STAR CCM+ software

Fig 5.4 Physics Models Selected for CFD Analysis of IRSS Systems

In Simcenter STAR CCM+, three basic approaches to modeling turbulence are available. These approaches are as follows:-

• Models that provide closure of the Reynolds-Averaged Navier-Stokes (RANS) equations such as K-Epsilon, K-Omega, RST and SAT

• Large eddy simulation (LES)

• Detached eddy simulation (DES)

It is observed from literature that different turbulence models are used for numerical simulation of exhaust flow in ejector-diffuser devices. The *use of Realizable k-Ɛ turbulence model is selected* because its performance parameters matched closer with experimental results compared to standard k-Ɛ. The Realizable k-Ɛ model is a two-equation model in which transport equations are solved for the turbulent kinetic energy k and its dissipation rate Ɛ. Also, a critical coefficient of the model, C_μ, is expressed as a function of mean flow and turbulence properties which is consistent with experimental observations in boundary layers. On the other hand, standard k-Ɛ model assumes C_μ as a constant.

5.5.3 Discretization scheme

Discretization scheme is employed when the face values between two adjacent volumes is required to be calculated by implementation of interpolation from the cell center values. Table 5.2 lists several upwind schemes provided in STAR CCM+ for RANS based turbulence models.

<u>**Table 5.2 Discretization schemes available for RANS models in FLUENT**</u>

Discretization scheme	Features
First-order upwind	Assumes that the face value equals to the upstream cell-cantered value
Second-order upwind	The face value is calculated through a Taylor series expansion of the upwind cell-cantered value and gradient

	between cells
Power law	For convection dominated flow, the scheme is essentially equivalent to the first-order scheme
QUICK	For structured grids aligned with the flow direction, it can provide more accurate performance; for unstructured or hybrid grids, it equals to second-order scheme
Third-order MUSCL	Applicable to all types of meshes and has a potential to increase accuracy for complicated three-dimensional flows

Based on literature review, it is observed that the first-order scheme is known to increase numerical diffusion especially when the flow is complex or is not aligned with the grid such as triangular and tetrahedral grids. It is only used for initial simulations. The accuracy achieved by power law scheme is normally equal to that of the first-order scheme and the accuracy generated by the QUICK scheme does not present important improvements over the second-order scheme.

In this research, all the simulations are based on the second-order upwind scheme.

5.5.4 Pressure-Velocity Coupling

Pressure-velocity coupling is only available for the segregated solver. For the current study, four common types of algorithms can be chosen in STAR CCM+ viz., SIMPLE (Semi Implicit Pressure Linked Equation), SIMPLER (SIMPLE Revised), SIMPLEC (SIMPLE Consistent) and PISO. SIMPLE or SIMPLER and SIMPLEC are usually used for steady-state flows, while PISO is recommended for transient flows. Convergence speed of solving uncomplicated flows (e.g. laminar flows) can be accelerated by using SIMPLEC rather than SIMPLE. For complicated flows involving turbulence, similar convergence rates will be reached with implementation of SIMPLE and SIMPLEC.

In this research, the SIMPLE algorithm is used in all simulations.

5.5.5 Near Wall Treatment

The presence of walls significantly affects the performance of turbulent flows. In the near wall region, the mean velocity field is affected due to no-slip condition at the wall. In the research, Prism layers are generated to capture the physics near the wall for all case studies.

5.5.6 Boundary Conditions

For all the case studies, the inlet boundary condition is defined for mass flow rate and temperature at the inlet of Nozzle as provided by the engine supplier. This also helps during experimental validation since mass flow and temperature at inlet to IRSS system for various loading conditions of the engine can be measured using suitable instrumentation. Exhaust duct wall is defined as thermal conducting baffle and for which thermal resistance is defined. The specified boundary conditions of the CFD models are shown in Figure 5.5. The turbulence parameters at the nozzle inlet are specified in terms of turbulence intensity and length scale. Turbulent intensity is calculated numerically using the value of velocity while turbulence length scale is taken as 7% of the hydraulic diameter of the geometry. Domain wall is defined as pressure boundary condition.

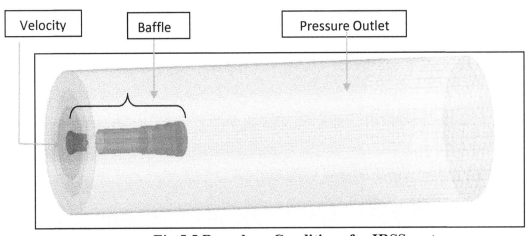

Fig 5.5 Boundary Conditions for IRSS system

5.5.7 Solution Convergence Study

A convergence study is performed for Case Study 2, to assess the convergence of CFD solutions to an acceptable level by monitoring the residuals of the conserved variables. Seven physical parameters are monitored during convergence namely continuity, X-momentum, Y-momentum, Z- momentum, energy, turbulence kinetic energy and turbulence dissipation rate. It was found from the convergence study that if the residuals reached upto the level of 10^{-4} for Continuity and turbulence Dissipation Rate (TDR) and10^{-5} for balance parameters and there is no significant change in the values of the residuals then the solution is considered as converged as shown in Fig 5.6.

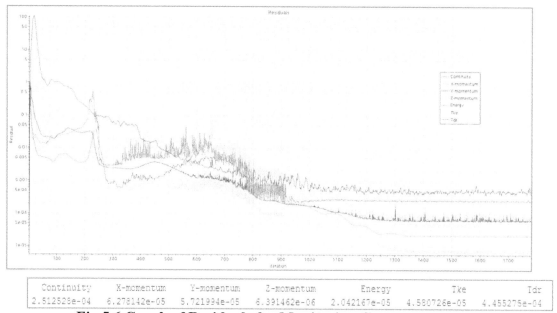

Continuity	X-momentum	Y-momentum	Z-momentum	Energy	Tke	Tdr
2.512529e-04	6.278142e-05	5.721994e-05	6.391462e-06	2.042167e-05	4.580726e-05	4.455275e-04

Fig 5.6 Graph of Residuals for Monitoring Convergence

5.6   FORMULATION OF CASE STUDIES

To understand the impact of change in key dimensional parameters of Nozzle, Mixing Tube and Diffuser on the overall infrared signature reduction levels and to compare it with exhaust system without IRSS, a total of 12 case studies are formulated as shown in Table 3. In all the cases, the standoff distance is maintained same at 220 mm.

Table 5.3 Formulation of Case Studies for Numerical Simulation

Case Study	Component Design Study	Integrated Ejector-Diffuser Configuration
1.	Ejector design - Nozzle exit area	• Round Nozzle • **RN 1 ($A_{ni}/A_{ne} = 1.1$)** • Single mixing tube • Four rings diffuser
2	Ejector design - Nozzle exit area and shape	• Round Nozzle • **RN 2 ($A_{ni}/A_{ne} = 2.1$)** • Single mixing tube • Four rings diffuser
3	Ejector design - Nozzle exit area	• Round Nozzle • **RN 3 ($A_{ni}/A_{ne} = 3.1$)** • Single mixing tube • Four rings diffuser
4	Ejector design - Nozzle exit shape	• **3 Lobed Nozzle** • Single cylindrical MT • Four ring diffuser
5	Ejector design - Nozzle exit shape	• **4 Lobed Nozzle** • Single cylindrical MT • Four ring diffuser
6	Ejector design - Nozzle exit shape	• **6 Lobed Nozzle** • Single cylindrical MT • Four ring diffuser
7	Ejector design - Nozzle exit shape	• **Elliptical Nozzle** • Rectangular mixing tube • Four ring diffuser
8	Mixing tube design	• **Twin Cylindrical MT** • **Standoff (0.5S+0.5S)** • Round Nozzle (RN 2) • Four ring diffuser
9	Diffuser design - Number of rings	• **Two Ring Cylindrical** • **Overlap- 1 Ring gap (1GP)** • Round Nozzle (RN 2) • Single cylindrical MT
10	Diffuser design - Ring overlap	• **Zero overlap (0RG)** • Two ring diffuser • Round Nozzle (RN 2) • Single cylindrical MT
11	Diffuser design - Ring overlap	• **Overlap- 2 Ring Gap (2RG)** • Two ring diffuser • Round Nozzle (RN 2) • Single cylindrical MT
12	Baseline infrared signature of non-stealth ship (Non-IRSS)	• **Cylindrical duct** • Uniform cross section

In addition to study of nozzle exit area, Case Study 2 is used for study of mixing tube design as single mixing tube case and also diffuser design as four rings diffuser. Similarly, in addition to study of diffuser design (number of rings), Case study 9 is used for study of diffuser design with one ring overlap.

5.7   CASE STUDY 1 - ROUND NOZZLE RN 1 (A_{ni}/A_{ne} = 1.1)

The case study RN1 represents a near cylindrical ejector having nozzle inlet diameter of 318 mm and exit diameter of 315 mm (A_{ni}/A_{ne} of 1.1). For this reason, it is also referred to as cylindrical nozzle. The dimensional design of integrated ejector-diffuser and the associated 3 D model are shown in Fig 5.7. The mixing tube inlet diameter is maintained at 0.336 mm. Thus, the nozzle exit area ratio with mixing tube (A_{mt}/A_{ne}) also varies and is 1.14. Exhaust gas flow characteristics are studied in terms of IR signature parameters and mass entrainment.

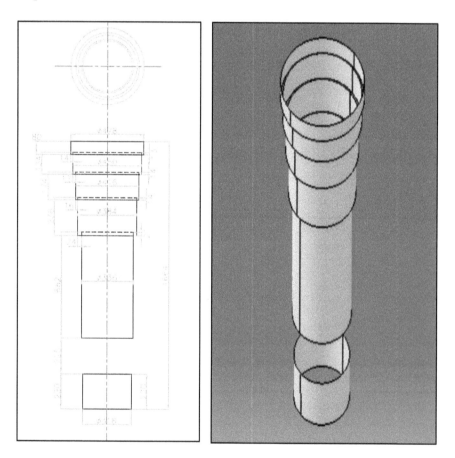

Fig 5.7 Dimensional Design and 3 D Model – Case Study 1

5.7.1   Exhaust Plume

The velocity and thermal profile along the length of IRSS are shown in Fig. 5.8. The thermal and velocity profile at device exit are shown in Fig 5.9. Salient observations are as follows:-

- Velocity of exhaust plume reaches a maximum value of **22 m/s in the mixing tube** and has a peak value of **20 m/s at the exit of diffuser** indicating minimal influence of ejector with area ratios closer to unity

- The core region of the plume largely remains unaffected as it progresses towards the funnel exit.

- Bulk mean temperature at exit is about **226 °C** which results in about **42.48% reduction in radiant intensity** or infrared signature due to plume.

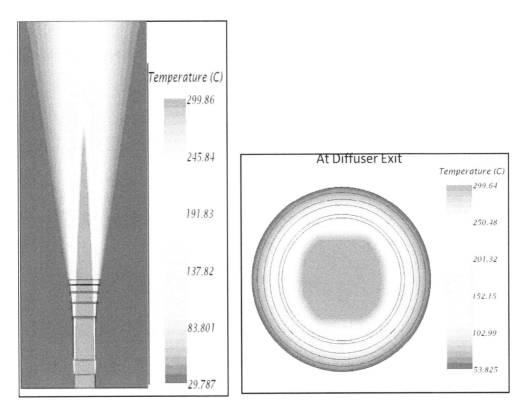

Fig. 5.8 Thermal Profile – Case Study 1

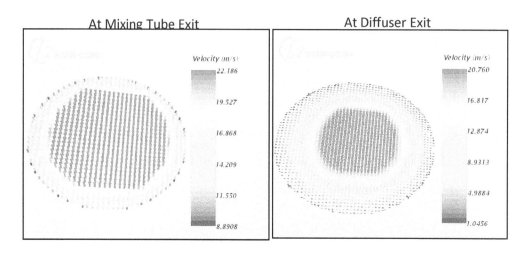

Fig 5.9 Velocity Profiles - Case Study 1

5.7.2 Diffuser Ring Duct Surface Temperatures

Diffuser ring wall surface temperatures of the four rings are shown Fig 5.10 from the bottom ring (R1) to top ring (R4).

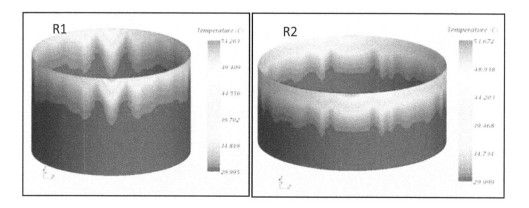

Ring No	Ring 1	Ring 2	Ring 3	Ring 4 (Top)
Avg. Temperature	34 °C	34.4 °C	37 °C	38.5 °C

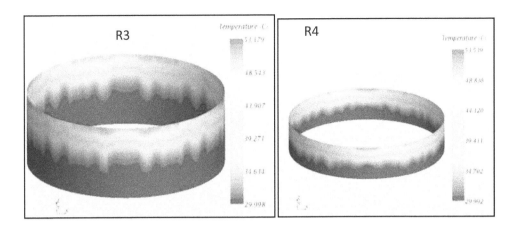

Fig 5.10 Diffuser Ring Wall Surface Temperatures - Case Study 1

5.7.3 Pressure Profile

The pressure profiles at inlet and exit of the nozzle are shown in Fig 5.11. The back pressure induced by the combination of mixing tube and diffuser is about 0.3 Pa while it is 33 Pa at nozzle inlet. Due to the near cylindrical shape of the ejector, the back pressure induced on the engine is very low compared to allowable pressure.

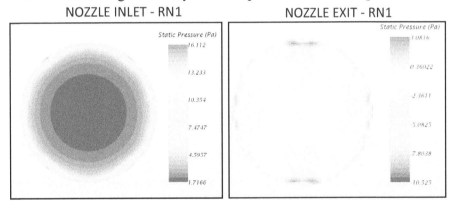

Fig 5.11 Static Pressure Profile at Inlet and Exit of Nozzle – Case Study 1

5.8   CASE STUDY 2 – ROUND NOZZLE RN 2 (A_{ni}/A_{ne} = 2.1)

The integrated ejector-diffuser for case study 2 (RN2) has a convergent nozzle with inlet diameter of 318 mm and exit diameter of 220 mm (A_{ni}/A_{ne} = 2.1). Consequently, the area ratio with mixing tube inlet also changes (A_{mt}/A_{ne}) as 2.23, which is twice that of case study 1. The dimensional model and 3 D model are shown in Fig 5.12. Exhaust gas flow characteristics are studied in terms of IR signature parameters and mass entrainment.

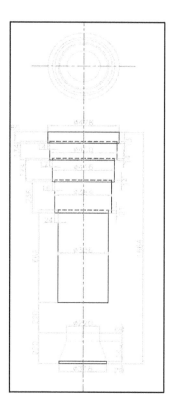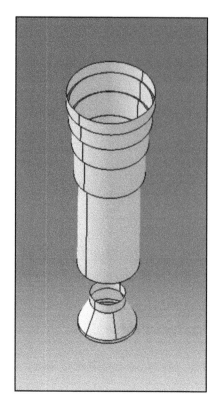

Fig 5.12 Dimensional Design and 3 D Model of Round Nozzle (RN2)

5.8.1   Exhaust Plume

Velocity and thermal profile along the length of IRSS are shown in Fig. 5.13. The thermal and velocity profile at exit in Fig 5.14. Salient observations are as follows:-

- Velocity of exhaust plume reaches a maximum value of **50 m/s in the mixing tube** and has a peak value of **43 m/s at the exit of diffuser**

- Core region of the plume decreases as it progresses towards device exit and temperature profile at exit indicates incomplete mixing within IRSS height

- Bulk mean temperature at exit is about **160 °C** which results in about **67.39% reduction in radiant intensity** or infrared signature due to plume.

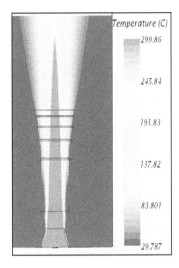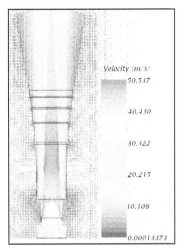

Fig. 5.13 Thermal and Velocity Profile Along IRSS - Case Study 2

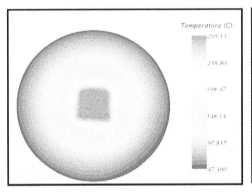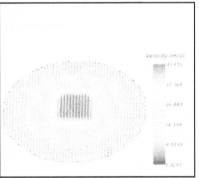

Fig 5.14 Thermal & Velocity Profiles at Device Exit – Case Study 2

5.8.2 Diffuser Ring Duct Surface Temperatures

Diffuser ring wall surface temperatures of the four rings are shown Fig 5.15 from bottom ring to top. The diffuser ring design based on film length enabled shielding more than 50% of the diffuser ring walls effectively from the high exhaust plume temperatures as seen from the blue region and the mathematical model for ring lengths enabled keeping the peak temperatures to less than $T_{amb}+30$ °C and the surface average ring temperatures to about **7 °C above ambient**.

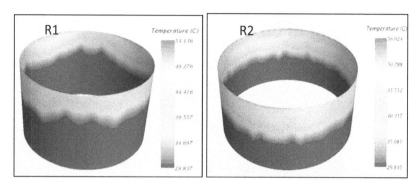

Ring No	Ring 1	Ring 2	Ring 3	Ring 4 (Top)
Avg. Temp.	33.3 °C	36.7 °C	35.3 °C	35.5 °C

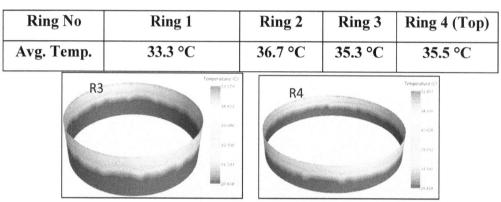

Fig 5.15 Diffuser Ring Wall Surface Temperatures for Case Study 2

5.8.3 Pressure Profile

The pressure profiles at inlet and along the length of exhaust stack are shown in Fig 5.16. Compared to the static pressure profile at inlet of exhaust stack without IRSS, it is seen that the average static pressure at inlet is considerably higher. The minimum and maximum pressures are about **445 Pa and 522 Pa with an average of 480 Pa**.

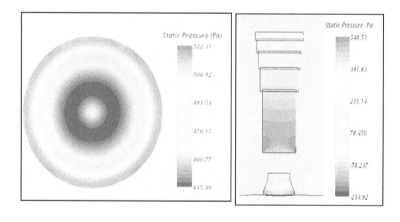

Fig 5.16 Static Pressure Profile in IRSS for Case Study 2

5.9   CASE STUDY 3 – ROUND NOZZLE RN 3 (A_{ni}/A_{ne} = 3.1)

The integrated ejector-diffuser for case study 3 (RN3) has a convergent nozzle with an inlet diameter of 318 mm and exit diameter of 180 mm (A_{ni}/A_{ne}) =3.1. Area ratio of mixing tube inlet to nozzle exit (A_{mt}/A_{ne}) is 3.5, which is nearly thrice that of case study 1. Exhaust gas flow characteristics are studied in terms of IR signature parameters and mass entrainment.

5.9.1   Exhaust Plume

The thermal profiles at mixing tube and diffuser exit are shown in Fig. 5.17. The velocity profiles at mixing tube and diffuser exit are shown in Fig 5.18. Salient observations are as follows:-

- Velocity of exhaust plume reaches a maximum value of **68 m/s in the mixing tube** and has a peak value of **63 m/s at the exit of diffuser**

- Core region of the plume decreases as it progresses towards device exit

- Bulk mean temperature at mixing tube exit is **196 °C** and at diffuser exit is **129 °C** which results in about **75.77% reduction in radiant intensity** or infrared signature due to plume.

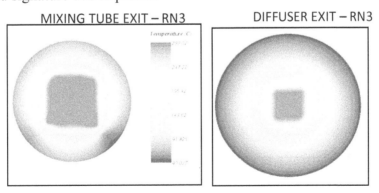

Fig. 5.17 Thermal and Velocity Profile Along IRSS - Case Study 3

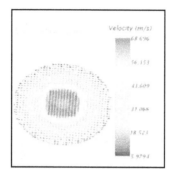 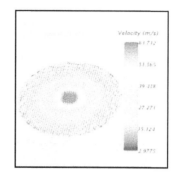

Fig 5.18 Thermal & Velocity Profiles at Device Exit – Case Study 3

5.9.2 Diffuser Ring Duct Surface Temperatures

Diffuser ring wall surface temperatures of case studies 3 are shown Fig 5.19. The low exit ratio has entrained considerably more air resulting in the mixing tube having longer film and peak temperatures are lower, thus providing higher view angle.

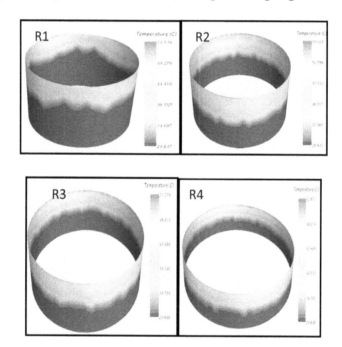

Fig 5.19 Diffuser Ring Wall Temperatures - Case Study 3

Ring No	Ring 1	Ring 2	Ring 3	Ring 4 (Top)
Avg. Temp.	35.5 °C	35.4 °C	36.7 °C	33.8 °C

5.9.3 Pressure Profile

The pressure profiles at inlet and exit of nozzle are shown in Fig 5.20. The exhaust back pressure due to mixing tube and diffuser together is very low compared to the static pressure profile at inlet of nozzle. Thus, the design of exit diameter of nozzle can be considered based on the back pressure allowance provided by the engine supplier. It is seen that the average static pressure at inlet is very high, beyond acceptable limits for gas turbine as specified in Table 2.1. The minimum and maximum pressures are about **1111 Pa and 1356.2 Pa with an average of 1215 Pa**.

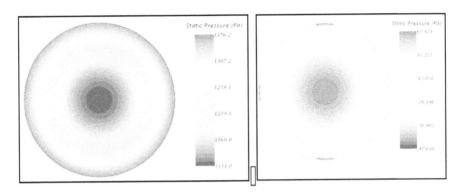

Fig 5.20 Static Pressure Profile at Nozzle Inlet – Case Study 3

5.10   CASE STUDY 4 – THREE LOBED NOZZLE (3LN)

The round nozzle is replaced with a three lobed nozzle while keeping the standoff, mixing tube and four ring diffuser same as that of baseline IRSS of Case 2.   The dimensional design and 3 D model are shown in Fig 5.21. The thermal, velocity and pressure profiles associated with the exhaust plume and duct surfaces obtained through simulation are presented below.

5.10.1 Exhaust Plume

The thermal and velocity profile along the length of IRSS are shown in Fig. 5.22 and at device exit are shown in Fig 5.23. Salient observations are as follows:-

- Velocity of exhaust plume reaches a maximum value of **53 m/s in the mixing tube** and has a peak value of **42 m/s at the exit of diffuser**

- Bulk mean temperature at exit is about **154 °C** which results in about **69.16% reduction in radiant intensity** or infrared signature due to plume.

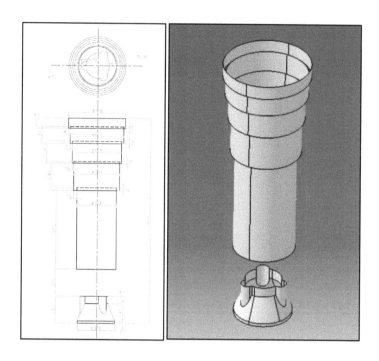

Fig 5.21 Dimensional Design and 3 D Model – Case Study 4

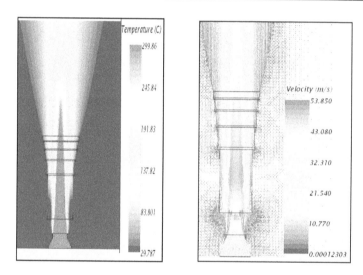

Fig. 5.22 Thermal and Velocity Profile – Case Study 4

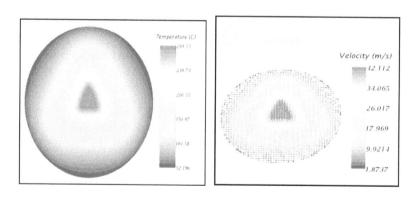

Fig 5.23 Thermal and Velocity Profile at Device Exit – Case Study 4

5.10.2 Diffuser Ring Duct Surface Temperatures

Diffuser ring wall surface temperatures of the four rings are shown Fig 5.24 from bottom ring to top. The thermal gradient along the length of each ring is higher. Though the surface average temperatures are about **10 °C above ambient**, the peak temperatures of the four rings are much higher ranging from **59 °C to 75 °C** (<T_{amb}+45).

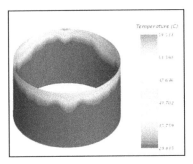 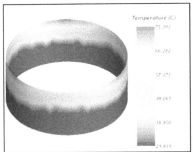

Ring No	Ring 1	Ring 2	Ring 3	Ring 4 (Top)
Avg. Temp	38.9 °C	38.3 °C	40.0 °C	32.6 °C

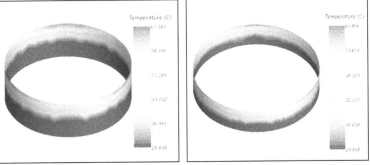

Fig 5.24 Diffuser Ring Wall Surface Temperatures – Case Study 4

5.10.3 Pressure Profile

Pressure profiles at inlet and along length of exhaust stack are shown in Fig 5.25. The min and max pressures are **403 Pa and 608 Pa with an average of about 496 Pa**.

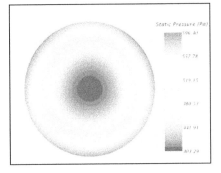 

Fig 5.25 Static Pressure Profile at Exit and along IRSS – Case Study 4

5.11   CASE STUDY 5 – FOUR LOBED NOZZLE (4LN)

The nozzle has four lobes while keeping the standoff, mixing tube and four ring diffuser same as that of baseline IRSS of Case 2. The dimensional design along with 3 D model is shown in Fig 5.26.

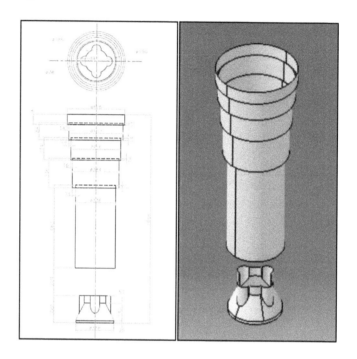

Fig 5.26 Dimensional Design and 3 D Model – Case Study 5

5.11.1 Exhaust Plume

The thermal and velocity profile along the length of IRSS are shown in Fig. 5.27 and thermal profile at device exit are shown in Fig 5.28. Salient observations are as follows:-

- Velocity of exhaust plume reaches a maximum value of **52 m/s in the mixing tube** and has a peak value of **41.6 m/s at diffuser exit**

- Bulk mean temperature at exit is about **152 °C** which results in about **69.74% reduction in radiant intensity** or infrared signature due to plume.

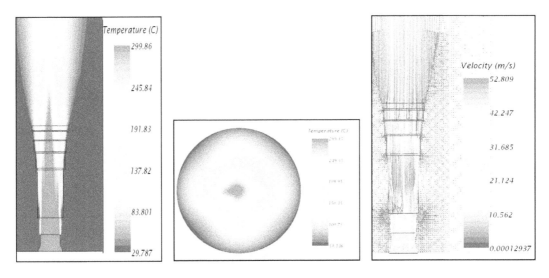

Fig. 5.27 Thermal and Velocity Profile – Case Study 5

5.11.2 Diffuser Ring Duct Surface Temperatures

Diffuser ring wall surface temperatures of the four rings are shown Fig 5.32 from bottom ring to top. The thermal gradient along the length of each ring is higher. For the same length of rings, peak temperatures are higher ranging from **61 °C to 78 °C** (**<T$_{amb}$+48**).

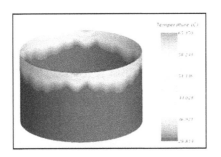 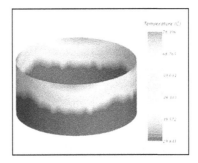

Ring No	Ring 1	Ring 2	Ring 3	Ring 4 (Top)
Avg. Temp	33.0 °C	41.7 °C	35.5 °C	38.8 °C

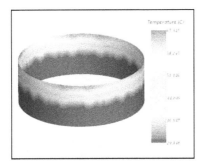 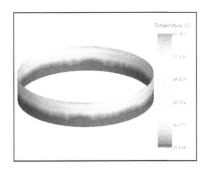

Fig 5.28 Diffuser Ring Wall Surface Temperatures – Case Study 5

5.11.3 Pressure Profile

The pressure profiles at inlet and along the length of exhaust stack are shown in Fig 5.29. The minimum and maximum pressures are about **390 Pa and 598 Pa with an average of about 492 Pa**.

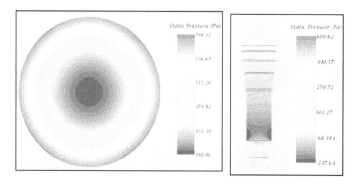

Fig 5.29 Static Pressure Profile at Exit and along IRSS – Case Study 5

5.12   CASE STUDY 6 – SIX LOBED NOZZLE (6LN)

The round nozzle is replaced with a six lobed nozzle while keeping the standoff, mixing tube and four ring diffuser same as that of IRSS of Case 2. The dimensional design is shown in Fig 5.30. The thermal, velocity and pressure profiles associated with the exhaust plume and duct surfaces obtained through simulation are presented below.

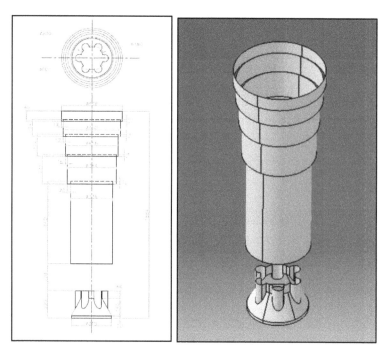

Fig 5.30 Dimensional Design and 3 D Model – Case Study 6

5.12.1 Exhaust Plume

The thermal and velocity profile along the length of IRSS are shown in Fig. 5.31 and at device exit are shown in Fig 5.36. Salient observations are as follows:-

- Velocity of exhaust plume reaches a maximum value of **53 m/s in the mixing tube** and has a peak value of **42 m/s at the diffuser exit**

- Bulk mean temperature at exit is about **153 °C** which results in about **69.45% reduction in radiant intensity** or infrared signature due to plume.

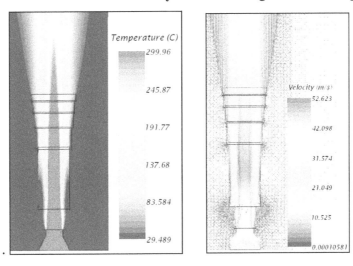

Fig. 5.31 Thermal and Velocity Profile – Case Study 6

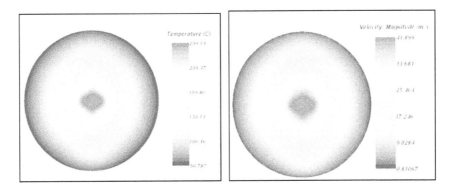

Fig 5.32 Thermal and Velocity Profile at Funnel exit – Case Study 6

5.12.2 Diffuser Ring Duct Surface Temperatures

Diffuser ring wall surface temperatures of the four rings are shown Fig 5.33 from bottom ring to top. The thermal gradient along the length of each ring is higher. Though the surface average temperatures are less than **10 °C above ambient** but the peak temperatures range from **59 °C to 79.7 °C (<T_{amb}+50)**.

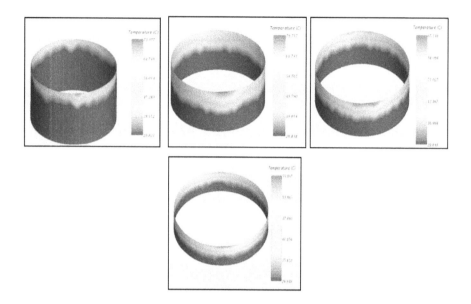

Ring No	Ring 1	Ring 2	Ring 3	Ring 4 (Top)
Avg. Temp	32.0 °C	40.0 °C	37.0 °C	37.0 °C

Fig 5.33 Diffuser Ring Wall Surface Temperatures – Case Study 6

5.12.3 Pressure Profile

The pressure profiles at inlet and along the length of exhaust stack are shown in Fig 5.34. The minimum and maximum pressures are about **395 Pa and 637 Pa with an average of about 505 Pa**.

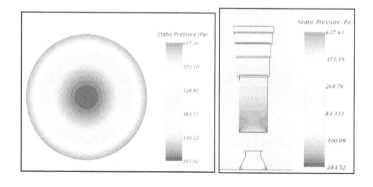

Fig 5.34 Static Pressure Profile at Exit and along IRSS – Case Study 6

5.13   CASE STUDY 7 – ELLIPTICAL NOZZLE (EN)

The case study has an elliptical nozzle and the mixing tube is provided with a skirt at the bottom to ensure that the exhaust gas does not flow outside the duct.   A two ring diffuser is provided.   The dimensional design and 3 D model are shown in Fig 5.35. The thermal, velocity and pressure profiles associated with the exhaust plume and duct surfaces obtained through simulation are presented below.

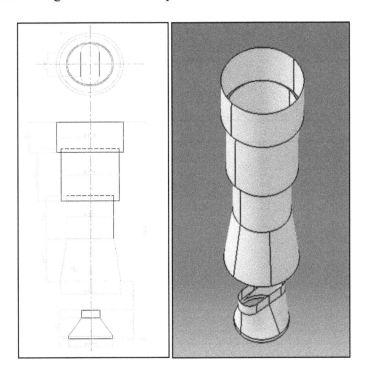

Fig 5.35 Dimensional Design and 3 D Model – Case Study 7

5.13.1 Exhaust Plume

The thermal and velocity profile along the length of IRSS are shown in Fig. 5.36 and at device exit are shown in Fig 5.37. Salient observations are as follows:-

- Velocity of exhaust plume reaches a maximum value of **45.7 m/s in the mixing tube** and has a peak value of **41.6 m/s at the exit of diffuser**

- Bulk mean temperature at exit is about **152.2 °C** which results in about **69.68% reduction in radiant intensity** or infrared signature due to plume.

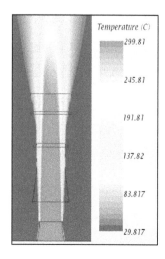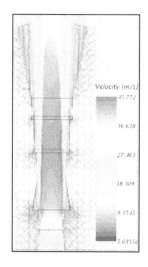

Fig. 5.36 Thermal and Velocity Profile – Case Study 7

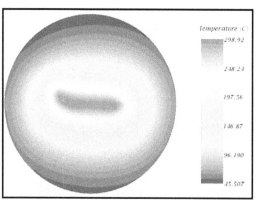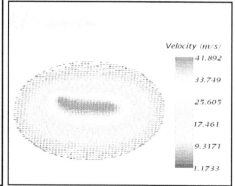

Fig 5.37 Thermal and Velocity Profile at Device Exit – Case Study 7

5.13.2 Diffuser Ring Duct Surface Temperatures

Diffuser ring wall surface temperatures of the two rings are shown Fig 5.38. The surface average temperatures of the two rings are **31.5 °C and 36.5°C** and peak temperatures are **53 °C to 62 °C (<T$_{amb}$+35).** Both surface average and peak temperatures are much lower than lobed nozzles.

108

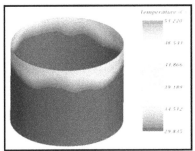 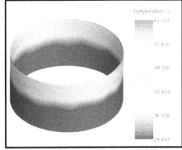

Ring No	Ring 1	Ring 2 (Top)
Avg. Temp	31.5 °C	36.5 °C

Fig 5.38 Diffuser Ring Wall Surface Temperatures – Case Study 7

5.13.3 Pressure Profile

The pressure profiles at inlet and along the length of exhaust stack are shown in Fig 5.39. The minimum and maximum pressures are about **385 Pa and 616 Pa with an average of about 465 Pa**.

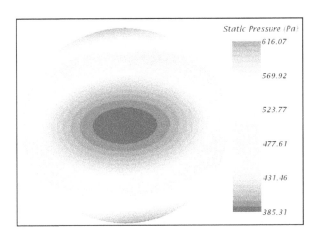 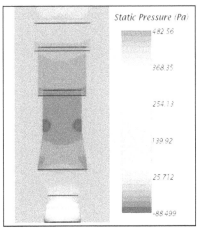

Fig 5.39 Static Pressure Profile at Exit and along IRSS – Case Study 7

5.14 CASE STUDY 8 – TWIN MIXING TUBE DESIGN (TMT)

The design of twin mixing tube has not been studied in the literature and is conceived as part of the research to explore the feasibility of including some part of mixing tube to increase the view angle protection (θ) offered by integrated ejector-diffuser technology (DRES Ball gives higher protection with back pressure penalty).

The integrated ejector-diffuser with round nozzle (RN2) and single mixing tube (SMT - 662 mm) is modified as case study 8. The total mixing tube length (662 mm) is divided into two equal parts (331 mm each). The total standoff distance of 220 mm is also divided equally to give 110 mm between nozzle and bottom mixing tube and 110 mm between bottom and top mixing tube. The area ratio between nozzle exit and mixing tube inlet is maintained same at 2.33. Dimensional design and 3 D model of Twin Mixing Tube is shown in Fig 5.40.

5.14.1 Exhaust Plume

The thermal and velocity profile along the length of IRSS are shown in Fig. 5.41 and at device exit are shown in Fig 5.42. Salient observations are as follows:-

- Velocity of exhaust plume reaches a maximum value of **51.3 m/s in the mixing tube** and has a peak value of **42.1 m/s at the exit of diffuser**

- Bulk mean temperature at exit is **155 °C**.

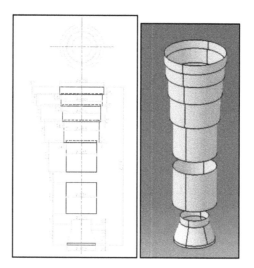

Fig 5.40 Dimensional Design and 3 D Model – Case Study 8

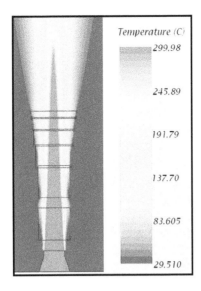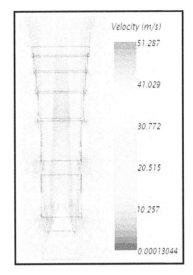

Fig. 5.41 Thermal and Velocity Profile along IRSS– Case Study 8

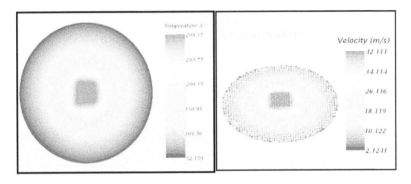

Fig 5.42 Thermal and Velocity Profile at Device Exit – Case Study 8

5.14.2 Diffuser Duct Surface Temperatures

Diffuser ring wall surface temperatures of the two rings are shown Fig 5.43. The peak and average surface average temperatures of the **bottom diffuser ring** are **60 °C and 51.75°C** respectively while that of **top diffuser ring** are **44.1 °C to 41.4 °C**. Both surface average and peak temperatures are much lower than lobed nozzles.

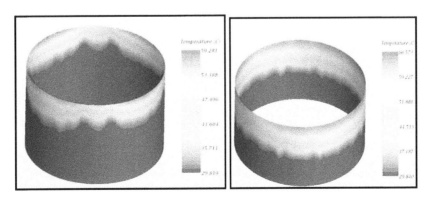

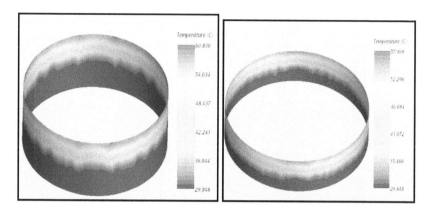

Ring No	Ring 1	Ring 2	Ring 3	Ring 4 (Top)
Avg. Temp	38.4 °C	38.0 °C	40.4 °C	33.7 °C

Fig 5.43 Diffuser Ring Wall Surface Temperatures – Case Study 8

5.14.3 Pressure Profile

The pressure profiles at inlet and along the length of exhaust stack are shown in Fig 5.44. The minimum and maximum pressures are about **390 Pa and 548 Pa with an average of about 468 Pa**.

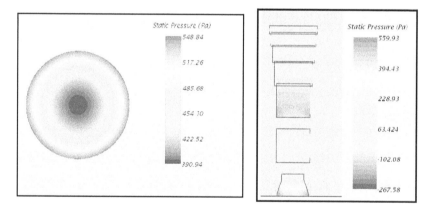

Fig 5.44 Static Pressure Profile at Exit and along IRSS – Case Study 8

5.15   CASE STUDY 9 – TWO RING DIFFUSER WITH ONE RING GAP (1RG)

To study the effect of number of rings and overlap between the rings of diffuser, Case Study 8 is taken up. It consists of round nozzle (RN2), single mixing tube (SMT) and a 2 ring diffuser. The overlap between mixing tube and bottom diffuser ring is kept

112

same as ring gap and hence, it is called 1RG. The dimensional design and 3 D model are shown in Fig 5.45.

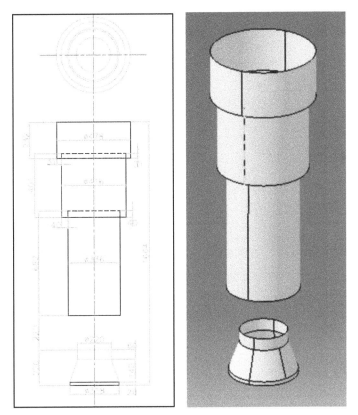

Fig 5.45 Thermal Profile at Diffuser Exit and along IRSS – Case Study 9

5.15.1 Plume Temperature Profile

The plume temperature profiles at mixing tube exit and diffuser exit are shown in Fig 5.46. Bulk mean temperatures are **192 °C** and **156 °C** respectively which is an indication of the extent of cooling due to entrainment through diffuser.

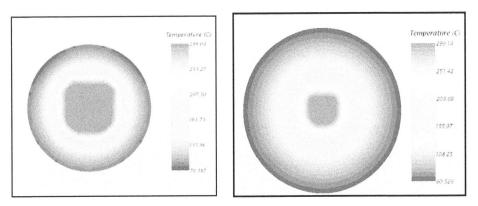

Fig 5.46 Thermal Profile at Mixing Tube Exit and Diffuser Exit – Case Study 9

5.15.2 Diffuser Ring Temperature Profiles

The thermal profile of the two diffuser ring with overlap equal to one ring gap is shown in Fig. 5.47. The surface average temperatures of bottom and top ring are **39 °C** and **40 °C** while their peak temperatures range from **65 °C to 70 °C (<Tamb+40 °C).**

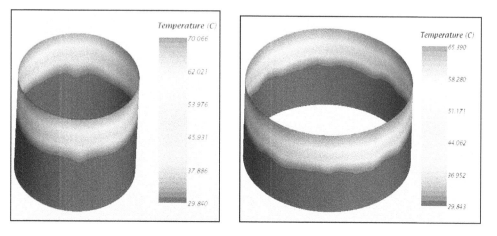

Fig 5.47 Thermal Profile of Bottom and Top Rings – Case Study 9

5.15.3 Pressure Profile

The max and min static pressure at center and near the walls are **395 Pa and 553 Pa** respectively as shown in Fig. 5.48 with an average of pressure loss of **474 Pa**. The pressure profile at diffuser inlet shows negative pressure indicating the cause of entrainment.

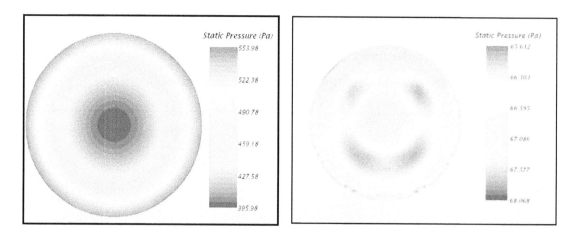

Fig 5.48 Pressure Profile at Nozzle Inlet and Diffuser Inlet – Case Study 9

5.16   CASE STUDY 10 – TWO RING DIFFUSER WITH NO OVERLAP (0RG)

The study is carried out with two ring diffuser because the ambient air film is likely to break when the length of ring is longer (compared to smaller lengths associated with four ring diffuser). The total length of rings and total gap is kept same as that of four ring diffuser. But there being no overlap, the overall length of diffuser becomes longer. Dimensional drawing and 3D model are shown in Fig 5.49.

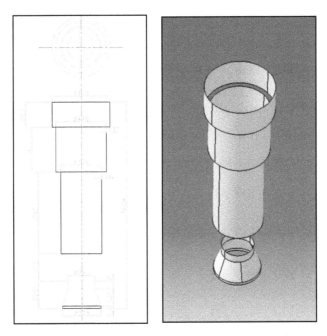

Fig 5.49 Dimensional Drawing & 3 D Model of Diffuser with No Overlap - Case Study 10

5.16.1 Plume Temperature Profile at Diffuser Exit

The plume temperature profiles at mixing tube exit and diffuser exit are shown in Fig 5.50. Bulk mean temperatures are **194 °C** and **156 °C** respectively which is similar to that of one ring overlap condition.

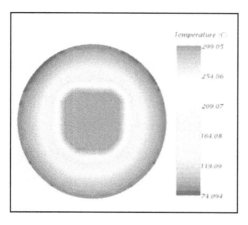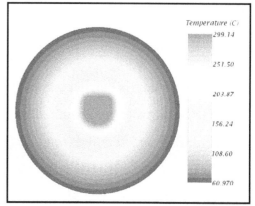

Fig 5.50 Thermal Profile at Mixing Tube Exit and Diffuser Exit – Case Study 10

5.16.2 Diffuser Ring Temperature Profiles

The thermal profile of the two diffuser rings is shown in Fig. 5.51. The peak temperatures range from **65 °C to 70 °C** while surface average temperatures of bottom and top ring are **39 °C** and **40 °C** respectively.

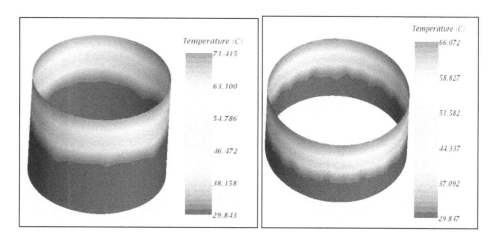

Fig 5.51 Thermal Profile of Bottom and Top Rings – Case Study 10

5.16.3 Pressure Profile

The max and min static pressure at center and near the walls are **396 Pa and 554 Pa** respectively as shown in Fig. 5.52 with an average of pressure loss of **474 Pa**. The pressure profile at diffuser inlet is about -50 Pa.

116

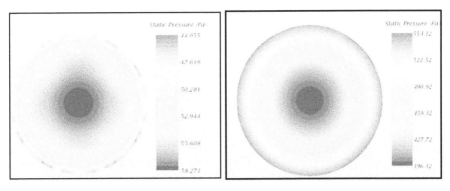

Fig 5.52 Pressure Profile at Nozzle Inlet and along IRSS for Case 9

5.17 CASE STUDY 11 – DIFFUSER WITH OVERLAP OF 2 RING GAP (2RG)

The doubling of overlap between the rings offers higher resistance to flow but is envisaged to guide the incoming ambient air for a longer length. The thermal and pressure profiles are presented below.

5.17.1 Plume Temperature Profile

The plume temperature profiles at Diffuser exit and along the length of IRSS are show in Fig 5.53. The bulk mean temperature is **156 °C**.

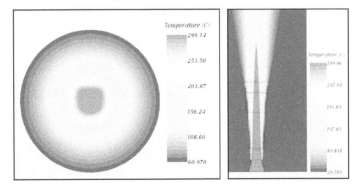

Fig 5.53 Thermal Profile at Diffuser Exit and along IRSS – Case Study 11

5.17.2 Diffuser Ring Temperature Profiles

The thermal profile of the two diffuser rings is shown in Fig. 5.54. The surface average temperatures of bottom and top ring are **38 °C and 38.5 °C** respectively while peak temperatures range from **66 °C to 71 °C (<Tamb+40 °C).**

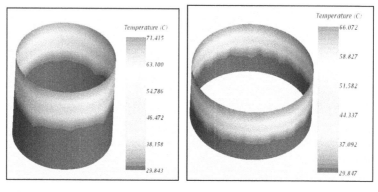

Fig 5.54 Thermal Profile of Bottom and Top Rings – Case Study 11

5.17.3 Pressure Profile

The max and min static pressure at center and near the walls are 396 Pa and 554 Pa respectively as shown in Fig. 5.55 with an average of pressure loss of 474 Pa.

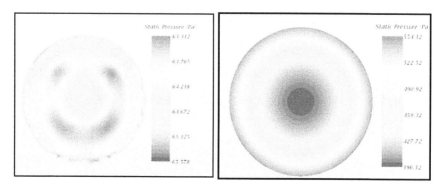

Fig 5.55 Pressure Profile at Nozzle Inlet and along IRSS – Case Study 11

5.18   CASE STUDY 12 - EXHAUST STACK WITHOUT IRSS (BASE LINE)

The exhaust stack of non-stealth ships consists of a duct of uniform cross section extending till the top of funnel exit as shown in Fig 5.56. The diameter of the stack is same as that at inlet of nozzle (318 mm) and total length is same as that of IRSS (1664 mm). Thermal insulation is often used on the outer side of the exhaust duct onboard non-stealth ships for safety of personnel working in the vicinity. However, the inner wall temperatures, which are seen by the IR homing system of the missile, are unaffected by the external insulation.

5.18.1 Exhaust Plume, Duct and Pressure Profiles

The velocity and thermal profile along the length of IRSS and at exit cross section for the exhaust duct without IRSS on non-stealth ships is shown in Fig. 5.56. It can be seen that flow is uniform with **minimal temperature drop and velocity along the length**. Thermal, velocity and pressure profiles at duct exit are uniform as shown in Fig. 5.57.

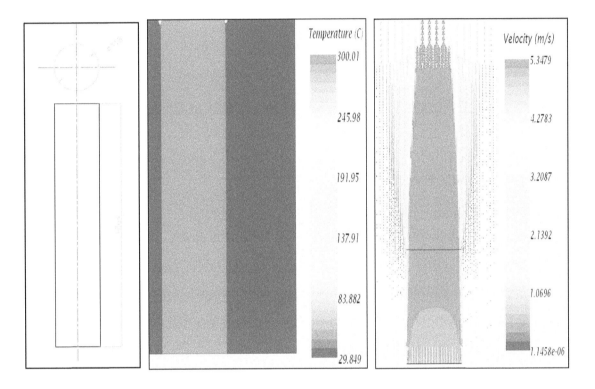

Fig 5.56 Thermal and Velocity Profile of Exhaust Plume - Case Study 12

The velocity of exhaust gas is due to engine alone and is as low as 5 m/s. There is negligible back pressure on the engine since straighter part of the duct is considered. Plume thermal profile at exit is uniform and same as that of inlet. Being considered here as baseline, infrared signature associated with the design is 100%.

119

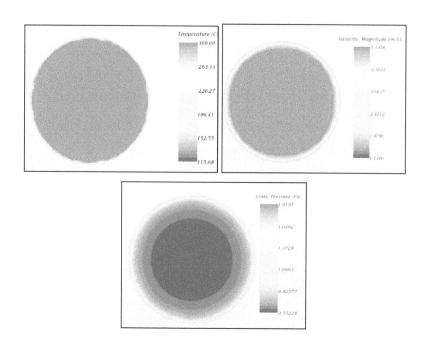

Fig 5.57 Thermal, Velocity & Pressure Profile at Duct exit – Case Study 12

5.19   CONCLUSION

The chapter provides a brief overview of various activities related to setting up of the software in terms of meshing, choosing suitable computational domain, grid independence study and the physics set up with a view to enable researches interested in the field to carryout similar simulations. Numerical simulation studies are carried out using Computational Fluid Dynamics software STAR CCM+ to identify the effect of component level design variations (03 nozzle exit areas, 05 nozzle shapes, 02 mixing tube designs, 02 diffuser ring designs, three overlap conditions) on the performance of an Integrated Ejector-Diffuser with respect to infrared suppression. Performance parameters of 12 case studies in terms of exhaust duct metal temperatures, plume temperatures, back pressure on engine etc ., are presented. Nozzle shape is observed to have considerable impact on secondary air but minimal effect on tertiary air induction. Nozzle convergence has significant effect on back pressure with simulations indicating that convergent nozzles with large area ratio ($A_{ni}/A_{ne} = 2.089$) have over 300 times higher average static pressure than cylindrical nozzle ($A_{ni}/A_{ne} = 1$). Total air entrained into diffuser is not significantly affected (<5%) by the number of rings (4 Ring and 2 Ring) subject to area ratio between diffuser exit and MT exit being same.

CHAPTER 6

EXPERIMENTAL STUDIES

6.1   INTRODUCTION

Fabrication of select case study models is taken up for experimental validation of the simulation results. The experimental studies on integrated ejector - diffuser IRSS systems are carried out using the Diesel Engine IRSS test facility at Naval Science and Technology Laboratory (NSTL), Visakhapatnam. The chapter presents the effect of component design on the performance of integrated IRSS system in terms of pumping ratios, fluid parameters at device exit, diffuser ring thermal profile and back pressure though measurement of the parameters at experimental setup.

6.2   CASE STUDIES CONSIDERED FOR EXPERIMENTAL VALIDATION

Experimental validation involves considerable expenditure in terms of financial resources, manpower and time. Computational simulation provides a cost effective, faster and low manpower intensive solution but the accuracy of estimate depends on the reasonableness of assumptions and extent of limitations in replicating real life variables. It is therefore essential that the results obtained through numerical simulation are validated with experimental measurements. For the current study, of the 11 case studies considered for simulation studies, fabrication is undertaken for four cases studies specifically selected to represent at least one component design. Five case studies were selected for experimental validation of component design changes observed through CFD and are as follows:-

(a) Nozzle design	- IRSS with round nozzle (RN1) (Case study 1)
	- IRSS with round nozzle (RN2) (Case study 2)
(b) MT design	- Twin Mixing Tube IRSS (Case study 8)
(c) Diffuser design	- Diffuser with 'One Ring gap overlap' (Case study 9)
	- Diffuser with No Ring overlap (Case study 10)

6.3   CASE STUDY 2 – ROUND NOZZLE (RN2) AND 4 RINGS DIFFUSER

6.3.1   Experimental Setup

The experimental set up for Case Study 2 is shown in Fig 6.1. It consists of round nozzle, standoff, single cylindrical mixing tube and a four ring diffuser. Thermocouple placed at nozzle exit provided feedback for maintaining the exhaust plume temperature through combustion motor cutting-in and cutting-off. Channels for placing thermocouples to record exhaust plume temperatures were positioned at nozzle exit and diffuser exit. An opening was provided on the mixing tube for inserting L type pitot tube.

Fig 6.1 Case Study 2- IRSS with Round Nozzle

6.3.2   Pressure Measurement

The measurement of static and dynamic pressure was undertaken using an L type pitot tube, details of which are provided in Chapter 3. The measurements were undertaken across the diameter at 9 locations for nozzle inlet, 10 locations for mixing tube inlet and

7 points at Diffuser exit. Measurement locations used for both pressure and temperature at Nozzle exit and Diffuser exit are shown in Fig 6.2.

Fig 6.2 Measuring Points at Nozzle and Diffuser Exit for Pressure & Temperature

The measured pressure data is shown in Table 6.2. It is observed that the static pressure is nearly constant across the plane while dynamic pressure is higher at points 2, 3 and 4 for nozzle and mixing tube indicating flow is comparatively higher towards that end. However, the dynamic pressure profile becomes uniform while reaching IRSS exit. The dynamic pressure is used to estimate the velocity and thereby, mass flow rate of primary fluid flow at inlet to nozzle, mixing tube and diffuser exit. The increase in flow is attributed to secondary and tertiary air induced through standoff and diffuser rings respectively.

Table 6.1 Measure Data of Static, Dynamic and Total Pressure for Case Study 2

Location	Nozzle Inlet (Pa)		Mixing Tube Inlet (Pa)		Diffuser/IRSS Exit (Pa)	
	Static	Dynamic	Static	Dynamic	Static	Dynamic
1	49.03	34.34	29.42	39.24	-	10.79
2	49.03	53.96	29.42	98.10	-	19.62
3	49.03	40.22	39.23	107.91	-	19.62
4	49.03	29.43	39.23	39.24	-	19.62
5	49.03	29.43	39.23	39.24	-	14.72
6	49.03	19.62	29.42	49.05	-	11.77
7	49.03	14.72	29.42	49.05	-	-
8	49.03	14.72	29.42	49.05	-	-
9	49.03	9.81	29.42	78.48	-	-
10	-	-	29.42	49.05	-	-

6.3.3   Exhaust Plume Temperature Measurement

Exhaust plume temperature is measured at nozzle exit and diffuser exit using a radiometric Infrared Imaging camera make FLIR Model T1020. The infrared images and corresponding temperature across diameter are shown in for nozzle exit in Fig 6.3 and diffuser exit in Fig 6.4. In Fig 6.3, the line probe shows lower values on the channel at places where holes are made for inserting the pitot tube and higher values at other locations. During first time measurement, thermocouple measurements are used to adjust the emissivity values in the IR camera so that channel temperature profile of line probe matches with absolute temperatures using thermocouples. Similar adjustment is also carried out with the channel at diffuser exit.

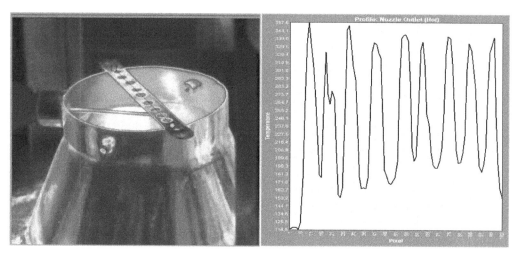

Fig 6.3 Temperature Profile at Nozzle Exit - Case Study 2

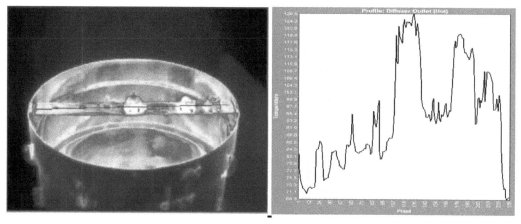

Fig 6.4 Temperature Profile at Diffuser Exit - Case Study 2

6.3.4 Exhaust Duct Temperature Measurement

The exhaust duct metal surface temperatures along the length of mixing tube and diffuser are shown in Fig 6.5 and Fig 6.6 respectively. The temperature of mixing tube increases from bottom to top because the cool air film formed due to ambient air induced through standoff breaks off. The diffuser rings have a cyclic temperature profile with temperatures increasing from bottom to top of each ring due to continued and sustained film formation as seen in Fig 6.6. The instantaneous peaks at the top end of each ring (except top ring) are attributed to the IR camera capturing temperature of hot plume through ring gaps.

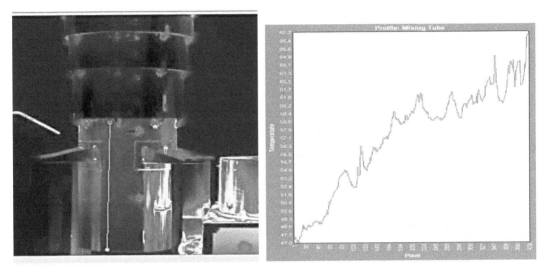

Fig 6.5 Temperature Profile of Mixing Tube - Case Study 2

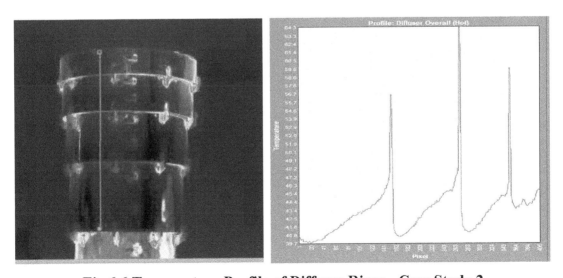

Fig 6.6 Temperature Profile of Diffuser Rings - Case Study 2

The temperature profiles of individual diffuser rings are shown in Fig 6.7.

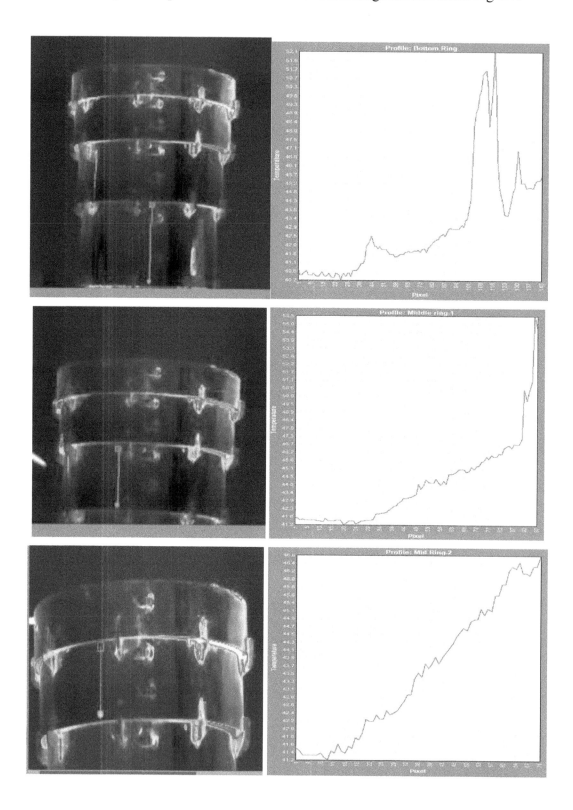

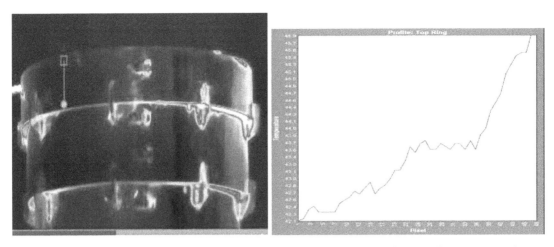

Fig 6.7 Temperature Profiles of Individual Diffuser Rings - Case Study 2

6.3.5   Tertiary Air through Diffuser Rings

Each of the four diffuser rings is supported by 8 brackets as shown in Fig 6.1. Velocity of ambient air induced through ring gaps is measured using an anemometer for each of the 8 brackets. The measured velocities at each bracket for each ring and the corresponding mass flow of air are shown in Table 6.3. In practice, it is not possible to measure the secondary air induced through standoff accurately. Hence, the mass flow of hot exhaust plume as primary fluid measured at nozzle exit and diffuser exit along with measured values tertiary air velocity at ring gaps is used to estimate secondary air induced into IRSS through standoff distance.

Table 6.2 Measurement of Tertiary induced into IRSS System- Case Study 2

	1	2	3	4	5	6	7	8
Measured Velocity of Ambient Air induced through Diffuser Ring Gaps								
Bottom	1	0.84	0.9	1.2	1.35	1.00	1	0.5
Middle 1	0.72	0.42	0.01	0.01	0.50	0.160	0.24	0.36
Middle 2	0.24	0.24	0.24	1.01	0.96	0.100	0.04	0.04
Top Ring	0.12	0.12	0.44	0.8	0.66	0.200	0.12	1.03
Calculated Mass Flow Rate of Ambient Air induced through Diffuser Ring Gaps								
Bottom	0.003947	0.003316	0.003553	0.004737	0.005329	0.003947	0.003947	0.001974
Middle 1	0.002105	0.001228	2.92E-05	2.92E-05	0.001462	0.000468	0.000702	0.001053
Middle 2	0.001057	0.001057	0.001057	0.004446	0.004226	0.00044	0.000176	0.000176
Top Ring	0.000231	0.000231	0.000849	0.001543	0.001273	0.000386	0.000231	0.001986

127

6.3.6 Remarks

The hot air generator generated 1 kg/s of hot plume at about 300 °C without the IRSS system. With the configuration of round nozzle and four ring diffuser, the mass flow at nozzle exit is recorded as 0.447 kg/s at an average temperature of 300 °C (with inherent variation of about 30 °C across diameter). The temperature during the trials varied by about 25 °C due to combustion motor cut-in and cut-off). The reduced mass flow is attributed to the back pressure induced by the IRSS system. Hence, simulation studies were carried out at measured inlet conditions for performance comparison.

6.4 CASE STUDY 1 – IRSS WITH CYLINDRICAL NOZZLE (RN1)

6.4.1 Experimental Setup

The experimental set up for Case Study 1 is shown in Fig 6.8. The round nozzle from case study 2 was replaced with a cylindrical nozzle. The standoff distance, single cylindrical mixing tube and a four ring diffuser are unchanged.

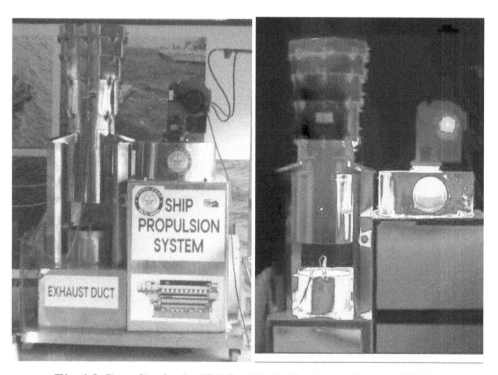

Fig 6.8 Case Study 1 - IRSS with Cylindrical Nozzle (RN1)

6.4.2 Pressure Measurement.

Measurement of static and dynamic pressure was undertaken using an L type pitot tube. The measurements were undertaken across the diameter at 11 points for cylindrical nozzle, 09 points for mixing tube and 7 points at Diffuser exit. Measured static pressure at nozzle inlet, dynamic pressure at nozzle exit and diffuser exit are in Table 6.4.

Table 6.3 Measure Data of Static and Dynamic Pressures for Case Study 3

Location	Nozzle Inlet (Pa)		Mixing Tube Inlet (Pa)		Diffuser/IRSS Exit (Pa)	
	Static	Dynamic	Static	Dynamic	Static	Dynamic
1	24.515	49.05	4.903	78.48	-	9.81
2	24.515	49.05	4.903	98.1	-	35.32
3	24.515	58.86	4.903	147.15	-	76.52
4	24.515	68.67	4.903	127.53	-	58.86
5	24.515	78.48	4.903	78.48	-	22.56
6	24.515	93.20	4.903	58.86	-	13.73
7	24.515	117.72	4.903	58.86	-	9.81
8	24.515	88.29	4.903	78.48	-	-
9	24.515	83.39	4.903	49.05	-	-
10	24.515	78.48	-	-	-	-
11	24.515	78.48				

6.4.3 Exhaust Plume Temperature Measurement

Infrared images using radiometric camera make FLIR model T1020 across nozzle exit and diffuser exit diameters are shown in Fig 6.9 and Fig 6.10 respectively. Temperatures at nozzle inlet post emissivity correction were at about 304 °C.

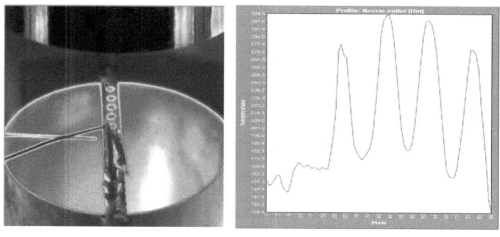

Fig 6.9 Temperature Profile at Nozzle Exit - Case Study 3

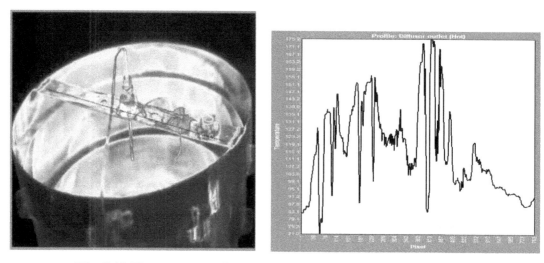

Fig 6.10 Temperature Profile at Diffuser Exit - Case Study 3

6.4.4 Exhaust Duct Temperature Measurement

The exhaust duct metal surface temperatures along the length of mixing tube and diffuser are shown in Fig 6.11 and Fig 6.12 respectively. The temperature of mixing tube increases from bottom to top because the cool air film formed due to ambient air induced through standoff breaks off. The diffuser rings have a cyclic temperature profile with temperatures increasing from bottom to top of each ring due to continued and sustained film formation as seen in Fig 6.6. The instantaneous peaks at the top of end of each ring (except top ring) are attributed to the IR camera capturing temperature of hot plume through the ring gaps.

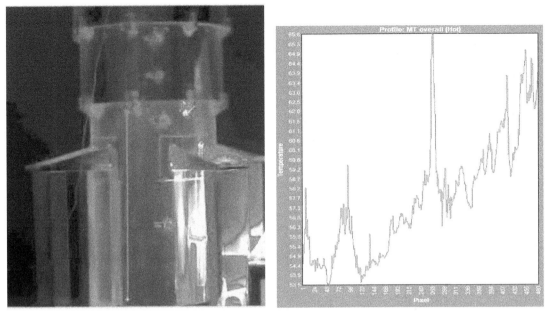

Fig 6.11 Temperature Profile of Mixing Tube - Case Study 3

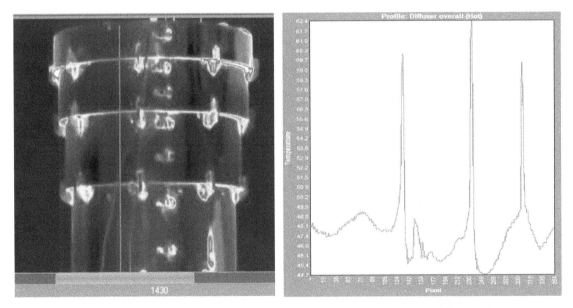

Fig 6.12 Temperature Profile of Diffuser Rings - Case Study 3

6.4.5   Tertiary Air through Diffuser Rings

Velocity of ambient air induced through ring gaps is measured using an anemometer for each of the 8 brackets. The measured velocities at each bracket for each ring and the corresponding mass flow of air are shown in Table 6.5.

131

Table 6.4 Measurement of Tertiary induced into IRSS System- Case Study 3

Measured Velocity of Ambient Air induced through Diffuser Ring Gaps –IRSS with Cylindrical Nozzle								
	1	**2**	**3**	**4**	**5**	**6**	**7**	**8**
Bottom	3.2	2.5	2.8	3.4	3.26	2.85	2.45	3.03
Middle 1	2.5	1.8	1.44	2.8	2.45	1.890	1.9	2.3
Middle 2	1.4	0.72	0.75	1.52	1.64	0.800	0.69	1.5
Top Ring	0.98	0.66	0.77	0.9	1.04	0.600	0.65	1.03
Calculated Mass Flow Rate of Ambient Air induced through Diffuser Ring Gaps								
Bottom	0.0140	0.0110	0.0123	0.0149	0.0143	0.0125	0.0107	0.0133
Middle 1	0.0081	0.0058	0.0047	0.0091	0.0080	0.0061	0.0062	0.0075
Middle 2	0.0068	0.0035	0.0037	0.0074	0.0080	0.0039	0.0034	0.0073
Top Ring	0.0021	0.0014	0.0016	0.0019	0.0022	0.0013	0.0014	0.0022

6.4.6   Remarks

The hot air generator generated 0.822 kg/s of hot plume at an average temperature of 225 °C without the IRSS system. The higher mass flow compared to round nozzle and lower mass flow compared to free flow (without IRSS) is attributed to the lower back pressure induced by the IRSS system with cylindrical nozzle. Hence, simulation studies were carried out at measured inlet conditions for performance comparison.

6.5   CASE STUDY 8 – IRSS WITH TWIN MIXING TUBE & 2 RING DIFFUSER

6.5.1   Experimental Setup

The experimental set up for Case Study 8 is shown in Fig 6.13. It aims to study the effect of design change to mixing tube by replacing the single mixing tube with a twin mixing tube of half the original length. The standoff distance of 220 mm is divided equally between nozzle, bottom MT and Top MT as shown in the dimensional drawing in Fig 4.12. Round nozzle and two ring diffuser (with one ring gap) complete the system. Thus, there is no change in total length of the system or in the dimensions of nozzle, diffuser and mixing tube. The performance parameters under consideration are the mass flow rate of secondary air through top and bottom mixing tubes and its effect on the thermal profile at IRSS exit. The air entrained through diffuser rings is also recorded to see if it is affected by the change in mixing tube and standoff combination.

Fig 6.13 Case Study 8 - IRSS with Twin Mixing Tube and 2 Ring Diffuser

6.5.2 Pressure Measurement

Measurement of static and dynamic pressure was undertaken using an L type pitot tube at 9 points across round nozzle inlet, bottom MT exit and diffuser exit. The measurements were taken across the diameter at 9 points on nozzle inlet, 10 points across bottom mixing tube exit and 7 points at Diffuser exit. Measured pressures are in Table 6.6.

Table 6.5 Measure Data of Static and Dynamic Pressures for Case Study 8

Location	Nozzle Inlet (Pa)		Bottom MT Exit (Pa)	Diffuser Exit (Pa)
	Static	Dynamic	Dynamic	Dynamic
1	29.420	19.62	19.62	4.905
2	34.323	24.525	34.335	9.81
3	29.420	34.335	39.24	9.81
4	29.420	34.335	49.05	19.62
5	39.227	39.24	49.05	117.72
6	49.033	39.24	98.1	88.29
7	49.033	39.24	78.48	9.81
8	49.033	34.335	117.72	-
9	39.227	34.335	39.24	-

10	-	-	29.43	-

6.5.3   Exhaust Plume Temperature Measurement

Infrared images using radiometric camera make FLIR Model T1020 across Bottom Mixing Tube and diffuser exit diameters are shown in Fig 6.14 and Fig 6.15 respectively.

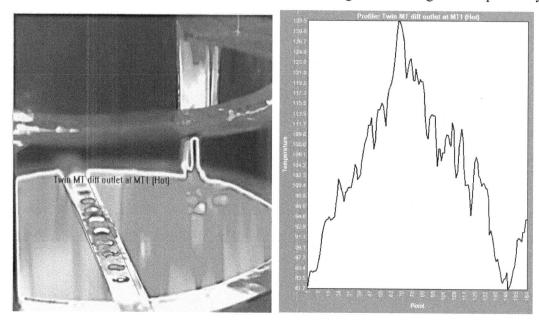

Fig 6.14 Temperature Profile at Bottom Mixing Tuber Exit - Case Study 8

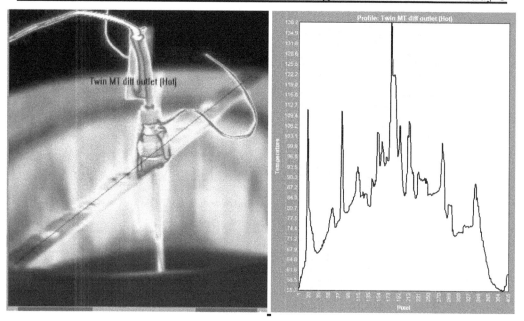

Fig 6.15 Temperature Profile at Two Ring Diffuser Exit - Case Study 8

6.5.4   Exhaust Duct Temperature Measurement

The exhaust duct metal surface temperatures along the length of bottom and top mixing tubes are shown in Fig 6.16 and the IR image of diffuser along with temperature profile is shown in Fig 6.17. It is observed that the top mixing tube is comparatively cooler than the bottom mixing tube.

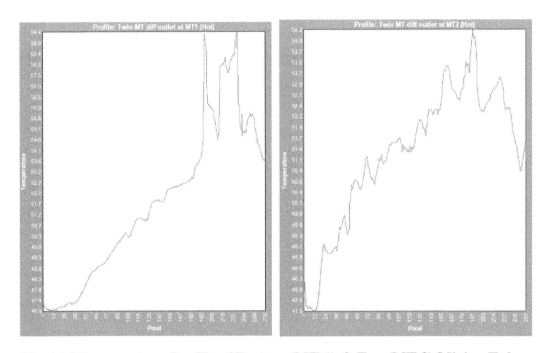

Fig 6.16 Temperature Profile of Bottom (MT 1) & Top (MT 2) Mixing Tubes

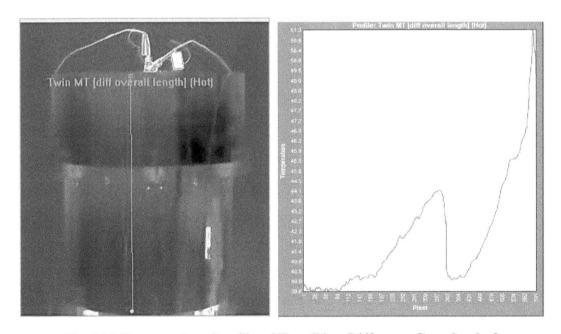

Fig 6.17 Temperature Profile of Two Ring Diffuser - Case Study 8

6.5.5   Tertiary Air through Diffuser Rings

Velocity of ambient air induced through ring gaps is measured using an anemometer for each of the 8 brackets. The measured velocities at each bracket for each ring and the corresponding mass flow of air are shown in Table 6.4.

Table 6.6 Measurement of Tertiary induced into IRSS System-Case Study 8

Measured Velocity of Ambient Air induced through Diffuser Ring Gaps						
	1	**2**	**3**	**4**	**5**	**6**
Bottom	2.5	2.53	2.7	2.08	2.85	1.950
Middle 1	1.62	2.13	2.01	1.91	2.45	1.400
Calculated Mass Flow Rate of Ambient Air induced through Diffuser Ring Gaps						
Bottom	0.023	0.023	0.024	0.019	0.026	0.018
Middle 1	0.014	0.018	0.017	0.016	0.020	0.012

6.5.6 Remarks

The hot air generator generated 0.514 kg/s of hot plume at an average temperature of 281 °C. The mass flow is comparable with round nozzle and four ring diffuser. Hence, simulation studies were carried out at measured inlet conditions for performance comparison.

6.6 CASE STUDY 9 & 10 – DIFFUSER RING OVERLAP STUDY

6.6.1 Experimental Setup. The case studies consist of IRSS with round nozzle, single mixing tube and a two ring diffuser assembly. The diffuser rings are concentrically separated by a gap and they also overlap with adjacent rings. To study the effect of ring overlap on the induction of tertiary air into IRSS and thereby the exhaust plume temperature at funnel exit, CFD simulations were carried out for 3 cases viz., no overlap, one ring gap overlap and two ring gap overlap. The ring gap between given set of rings and number of rings is kept the same for all three case studies. The experimental set up for Case Studies 9 and 10 are shown in Fig 6.18 and 6.19 respectively. Three sets of vertically separated holes are drilled at each bracket for Mixing Tube and Bottom Ring such that they correspond to nil overlap, one ring gap overlap and two ring gap overlap for a given set of rings. In Fig 6.18, one set of holes are seen on MT and bottom ring which implies that the ring is mounted in the holes that correspond to overlap equal to one ring gap. Two sets of holes are seen in Fig 6.19 indicating that there is nil overlap between the rings. The performance parameters under consideration are the mass flow rate of tertiary air through top and bottom diffuser rings, thermal profile of exhaust duct metal for diffuser and exhaust plume at IRSS exit.

Fig 6.18 One Ring Gap Overlap　　　**Fig 6.19 Nil Overlap Diffuser**

6.6.2　Tertiary Air through Diffuser Rings

The diffuser rings are supported on 6 brackets. Velocity of ambient air induced through ring gaps is measured using an anemometer for each bracket. The measured velocities at each bracket for each ring and the corresponding mass flow of air for IRSS with one ring overlap and nil overlap are shown in Table 6.5.

Table 6.7 Tertiary Air induced into IRSS System- Case Study 9 & 10

Gap Location	IRSS with One Ring Gap- Case Study 9						
MT-Bot (m/s)	2.00	2.00	1.98	1.40	1.45	1.35	1.70
Bot-Top (m/s)	1.10	1.30	1.10	0.96	1.07	0.90	1.07
MT-Bot (kg/s)	0.0178	0.0178	0.0176	0.0124	0.0129	0.0120	0.0904
Bot-Top (kg/s)	0.0090	0.0106	0.0090	0.0079	0.0088	0.0074	0.0526
	IRSS with Nil Gap- Case Study 10						
Bot-Top (m/s)	0.0188	0.0172	0.0191	0.0124	0.0178	0.0153	0.1007
MT-Bot (m/s)	0.0098	0.0086	0.0109	0.0092	0.0106	0.0117	0.0608
Bot-Top (kg/s)	2.12	1.94	2.15	1.4	2.01	1.72	1.89
MT-Bot (kg/s)	1.2	1.05	1.33	1.12	1.3	1.43	1.24

6.6.3 Exhaust Plume Temperature Measurement

Temperature vs distance graph obtained through infrared imaging across diffuser exit diameter for IRSS with and without overlap are shown in Fig 6.20 (a) and (b) respectively. The diametric length of 478 mm is divided into number of pixels depending upon the distance of obtaining image.

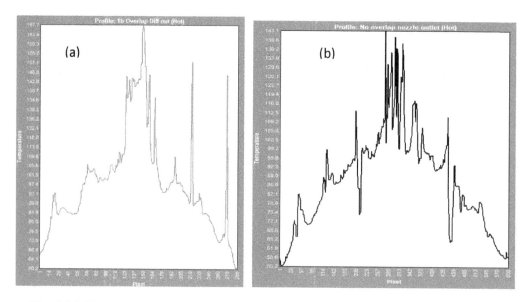

Fig 6.20 Temperature Profile at IRSS Exit for Diffuser (a) with and (b) without Overlap

6.6.4 Exhaust Duct Temperature Measurement

The exhaust duct metal surface temperatures along the length of diffuser rings for IRSS with and without overlap are shown in Fig 6.21 and Fig 6.22 respectively. The instantaneous peak near the boundary between the rings is due to exhaust plume temperature visible through gap. The duct metal temperatures of diffuser rings without overlap are cooler than those with overlap.

6.6.5 Remarks

The hot air generator generated 0.585 kg/s of hot plume at an average temperature of 300 °C. The reduction in mass flow is attributed to the back pressure induced by the IRSS system with round nozzle.

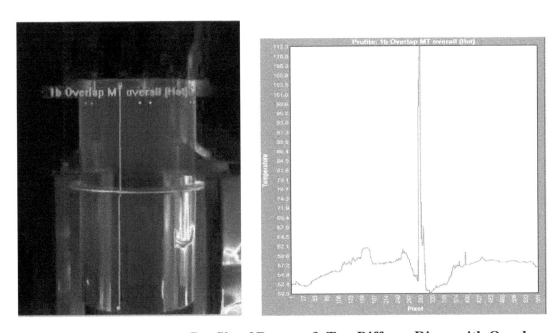

Fig 6.21 Temperature Profile of Bottom & Top Diffuser Rings with Overlap

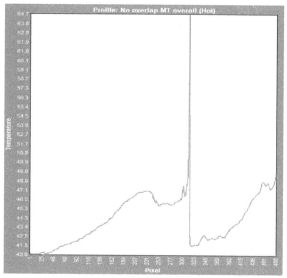

Fig 6.22 Temperature Profile of Bottom & Top Diffuser Rings without Overlap

6.7   Conclusion

The chapter presents experimental trials of six different IRSS systems carried out at the Diesel Engine IRSS test facility at Naval Science and Technology Laboratory (NSTL), Visakhapatnam. The collected data of exhaust plume temperatures, duct wall temperatures, static pressure and velocity of secondary air entrained for each trial is presented. Infrared imaging is undertaken and extracted data is presented. Experimental measurements have shown higher secondary air induction for the twin mixing tube design (88%) compared to simulation (68%). It is observed that TMT design has lower mass flow average plume temperatures at diffuser exit (151 °C) compared to SMT design (157 °C). A significant observation is that Integrated ejector-diffuser with 'No Overlap' emitted 9% higher infrared signature than 4 Ring Diffuser and 5% higher signature than 2 Ring Diffuser. This is attributed to the 15% increase in ring length over Diffuser with 2 Ring Gap. This highlights that optimal air entrainment does not always result in optimal IR suppression.

CHAPTER 7

RESULTS AND DISCUSSIONS

7.1   INTRODUCTION

The research work commenced with a study of mathematical formulations associated with design of subsystems/ components of passive IRSS system and developed two mathematical models, viz., 'Ejector-Diffuser for Ship Applications (EDSA)' for dimensional design and 'Aero-Thermal Performance Model (ATPM)' for preliminary performance assessment. Using EDSA and ATPM, alternate dimensional designs of integrated ejector-diffuser IRSS are generated for a given set of ship structural and fluid parameters. Numerical simulations studies are carried out for 12 case studies to study the effect of design variations of nozzle, mixing tube and diffuser on the infrared signature suppression achieved by the integrated device. Experimental tests on five full scale models are carried out at the Diesel Engine IRSS test facility at Naval Science and Technology Laboratory (NSTL), Visakhapatnam. Results of mathematical modeling, numerical simulation and experimental tests are analyzed to understand the effect of component/ subsystem design on the performance of integrated IRSS system in terms of pumping ratios, exhaust plume temperature at device exit, diffuser ring temperatures and back pressure imposed on engine. The effect of IRSS design on Missile Lock-On Range is highlighted.

7.2   GROUPING OF CASE STUDIES

The results of twelve case studies for which simulation studies have been carried out are grouped together into six groups as shown in Table 7.1. Groups A to D study the effect of design variations of nozzle, mixing tube and diffuser subsystems on the performance of integrated ejector-diffuser IRSS system. Group E compares the infrared signature suppression achieved by a ship with typical IRSS with that by straight duct on non-stealth.

Table 7.1 Grouping of Case Studies

Group	Area of Study	Case Study No.	Component under study	Unchanged Components
A	Nozzle Exit Area	1 to 3	• Round Nozzle 1 • Round Nozzle 2 • Round Nozzle 3	• Standoff • Single MT • 4 Rings Diffuser
B	Nozzle Exit Shape	2 & 4 to 7	• Round Nozzle (RN 2) • 3 Lobed Nozzle • 4 Lobed Nozzle • 6 Lobed Nozzle • Elliptical Nozzle	• Standoff • Single MT • 4 Ring Diffuser • Overlap = b* • Elliptical (2 ring diffuser)
D	Mixing tube design (MT & Standoff)	8 & 9	• Twin MT with Standoff (0.5S+0.5S) • Single MT	• Round nozzle • 2 ring diffuser • Overlap = b*
E	Number of diffuser rings	2 & 9	• Four Ring • Two Ring	• Round Nozzle • Standoff • Single MT • Overlap = b*
F	Ring overlap	9 to 11	• No overlap • Overlap = b* • Overlap = 2b*	• Round Nozzle • Standoff • Single MT • 2 ring diffuser
A	IR Signature of Stealth and Non-stealth ships	2 & 12	Uniform Duct Vs Baseline IRSS	• Duct inlet dia • Duct length • Plume properties

b* - Ring gap associated with adjacent rings under consideration

7.3 NOZZLE DESIGN STUDY - CASE STUDIES 1 to 7

Nozzle defines the driving force for inducing large amount of ambient air into hot exhaust plume and thus, causes reduction of ship's infrared signature due to exhaust system. There are three key performance criteria for design of nozzles viz., maximize plume velocity, maximize mixing and minimize back pressure on engine. The measurable parameters chosen for analyzing the effect of nozzle inlet-exit area ratio (RN1, RN2 and RN3) and exit shape (RN1, RN2, 3LN, 4LN, 6LN and EN) in meeting the component performance criteria are shown in Table 7.2.

Table 7.2 Parameters for Design Study of Nozzle Subsystem

Ser	Component Performance Criteria	Parameters for Analysis
(a)	Maximize exhaust plume velocity	(i) Primary mass flow along IRSS
		(ii) Pumping Ratio of secondary, tertiary air entrained
		(iii) Exhaust duct temperature profile
(b)	Minimize back pressure induced due to nozzle	(iv) Static pressure profile at nozzle inlet
(c)	Maximize mixing of hot plume with ambient air drawn into the system	(v) Exhaust plume temperature profile at IRSS exit

7.3.1 Primary Mass Flow Rate

Exhaust plume mass flow rate of 1 Kg/s at 300 °C at nozzle inlet is considered for all studies. Primary mass flow refers to the flow of gas and ambient air mixture through exhaust duct from inlet to exit. Nozzle (RN1) with circular shape at exit and inlet-exit area ratio close to unity is considered the baseline for estimating % variation. The primary mass flow rate along the length of IRSS system for nozzles of different shapes based on simulation studies is shown in Fig 7.1.

The mass flow of primary fluid at IRSS exit is higher by over 30% for convergent nozzles vis-à-vis RN1. Among nozzles of different shapes (with same A_{ni}/A_{ne}), 4 LN provides higher mass flow rate (38.13%) followed by 6LN (36.9%), EN (36.8%), 3LN (35%) and RN2 (31%). The % deviation from mean primary mass flow at IRSS exit for different nozzle shapes is less than 4% and therefore, it is inferred from simulation studies that design of nozzle subsystem in terms primary mass flow at IRSS exit is affected by nozzle area ratio A_{ni}/A_{ne} compared to nozzle shape.

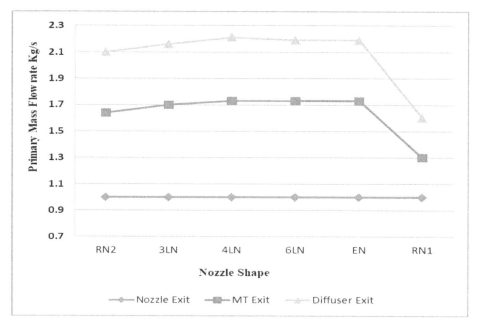

Fig 7.1 Primary Mass Flow Rate at Diffuser Exit for Different Nozzle Shapes

7.3.2 Pumping Ratio

Pumping ratio is a measure of quantity of secondary and tertiary air induced through standoff and diffuser rings respectively. It indicates the plume cooling effectiveness of an IRSS system. The pumping ratio of secondary air (Φ_s) induced through standoff, tertiary air (Φ_t) induced through diffuser rings, and total flow at IRSS exit vis-à-vis inlet (Φ_T) are studied. The simulated and measured pumping ratios (Φ_s, Φ_t & Φ_T) are shown in Fig 7.2.

The simulated pumping ratio of secondary air (Φ_s) induced through standoff for Case study 1 (RN1) is very less (10%) compared to convergent nozzles (64-73%). The effect of (A_{mt}/A_{ne}) is reduced when it comes to pumping ratio of tertiary air (Φ_t). The simulated tertiary and total pumping ratios for RN1 (Φ_t - 28% & Φ_T - 138 %) are considerably less than that of convergent nozzles (Φ_t - 46% & Φ_T - 210 %). Experimental measurements for RN1 (Φ_T - 145.6) % and RN (Φ_T - 199%) are in agreement with the trends observed through simulation. The considerably lower suction of ambient air with

RN1 is attributed to the lower value of mixing tube inlet to nozzle exit area ratio (A_{mt}/A_{ne}) of 1.057 while it is 1.52 for convergent nozzles.

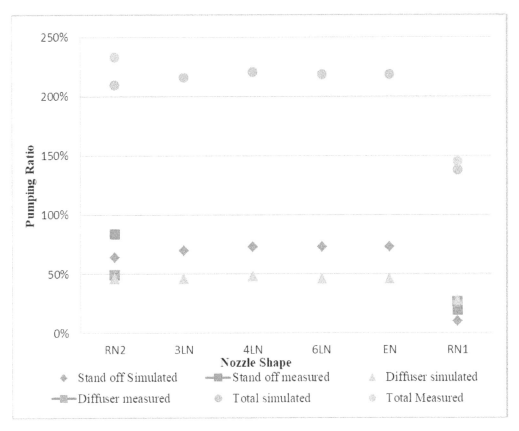

Fig 7.2 Simulated and Measured Pumping Ratios for different Nozzles

Case studies 2,4,5,6 and 7 have *same* mixing tube inlet to nozzle exit area ratio (A_{mt}/A_{ne}). Nozzle shape has marginal effect on the simulated pumping ratio through standoff (RN- 64%, 3LN - 70%; 4LN, 6LN and EN - 73%) but negligible effect on the pumping ratio of tertiary air (Φ_t) induced through diffuser rings (4LN - 48%; RN, 3LN and 6LN - 46%). The overall effect of nozzle shape on pumping ratio (Φ_T) varies from 210 to 221% with 4 lobed nozzle offering higher pumping ratio (RN- 210%, 3LN - 216%; 4LN-221%, 6LN- 219% and EN - 219%). The effect of nozzle shape on pumping ratio through standoff is attributed to the higher velocity of induced air at the position of lobes (~ 16 m/s for 4 LN) compared to RN (~7 m/s) as seen in Fig 7.3. As the plume travels along the length of IRSS, the effect of nozzle shape reduces and velocity profile at diffuser exit is more uniform as seen in Fig 7.4.

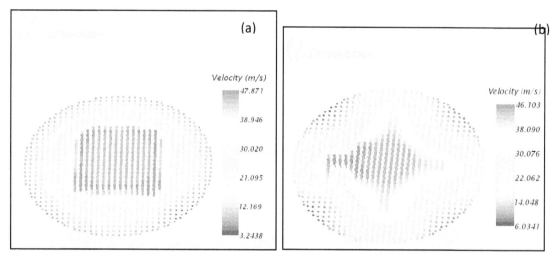

Fig 7.3 Velocity at MT Exit for (a) Round (Φ_s-64%) & (b) 4 Lobed Nozzles (Φ_s-73%)

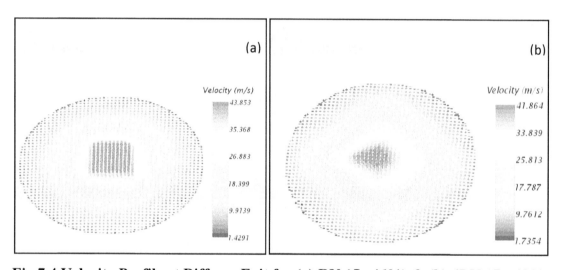

Fig 7.4 Velocity Profile at Diffuser Exit for (a) RN (Φ_t-46%) & (b) 4LN (Φt-48%)

Experimentally measured pumping ratio of secondary air (Φ_s) for round nozzle is higher (70%) than simulation (64%) while tertiary pumping ratio (Φ_t) is measured (13%) considerably less than simulation (46%). This compensates the error between simulation and measurement to a large extent and the total pumping ratios of IRSS (Φ_T) measured (199%) and simulated (201%) are in good agreement. For Cylindrical nozzle, the difference between simulated (Φ_s- 19%, Φ_t - 28% & Φ_T – 138 %) and experimental measurements (Φ_s- 10%, Φ_t - 26.4% & Φ_T – 145.6 %) is much less especially for tertiary and total pumping ratios.

It is, thus, observed that nozzle shape has marginal effect (11%) on the total pumping ratio (Φ_T) while mixing tube to nozzle exit area ratio (A_{mt}/A_{ne}) has significant effect.

7.3.3 Exhaust Duct Temperature

High temperature of diffuser rings is the prime source of ship infrared signature. The thermal profile along the length of diffuser rings and mixing tube is affected by the amount of secondary and tertiary air induced due to nozzle and thus, is a key driver in providing protection from detection by missile IR seekers. Typically, the exhaust duct is made of stainless steel 316L and is considered as a grey body with an emissivity of 0.9. The surface average diffuser wall temperatures obtained through simulation for case studies 2 to 7 are shown in Fig 7.5. It is observed that the diffuser wall temperatures for all the IRSS systems with cylindrical and convergent nozzles are near identical with no specific pattern.

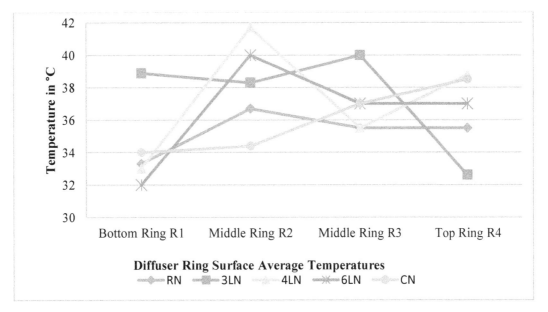

Fig 7.5 Simulated Surface Average Wall Temperatures of Diffuser Rings for Case Studies 2 to 6

Experimental measurements included IR imaging of diffuser rings and plotting the temperatures along the length. The diffuser section temperature profiles obtained from CFD simulation and IR imaging for case study 2 (RN) and case study 3 (RN1) are shown in Fig 7.6 and Fig 7.7 respectively.

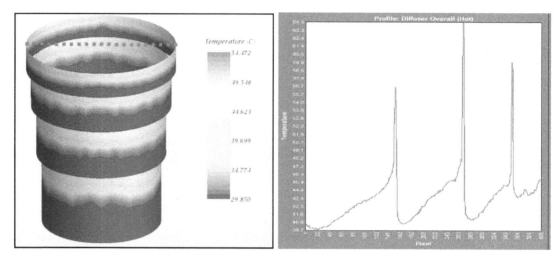

Fig. 7.6 Simulated Thermal Profile Vs Measured IR Image Plot – Case Study 2

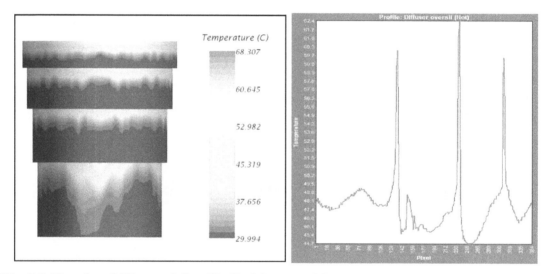

Fig 7.7 Simulated Thermal Profile Vs Measured IR Image Plot – Case Study 3

For the given lengths of diffuser rings, thermal reduction achieved using IRSS of different nozzle shapes has a good correlation between CFD simulation and experimental IR imaging. Wall temperatures associated with round nozzle (RN2) are lower than that of RN3.

7.3.4   Static Pressure Profile.   The static pressure profile along the length of IRSS system for cylindrical, round and lobed nozzles is shown in Fig 7.8. It is observed that the static pressure within the IRSS system is predominantly due to nozzle.

148

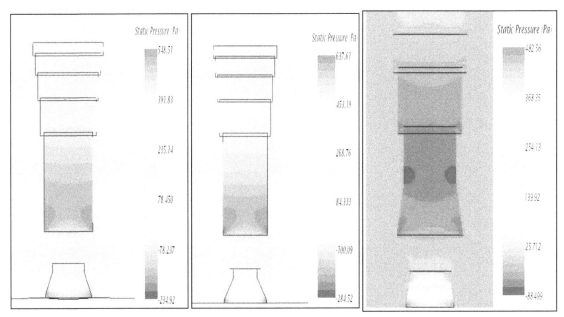

Fig 7.8 Pressure Profile along IRSS System for Round, Lobed and Elliptical Nozzles

The static pressure profiles across inlet to nozzle for cylindrical and lobed nozzle are shown in Fig. 7.9. Average static pressure for convergent nozzles with large area ratio (A_{ni}/A_{ne} = 2.089) is over 300 times higher than cylindrical nozzle (A_{ni}/A_{ne} = 1). Static pressure is therefore extremely sensitive to the area ratio of nozzle inlet to exit.

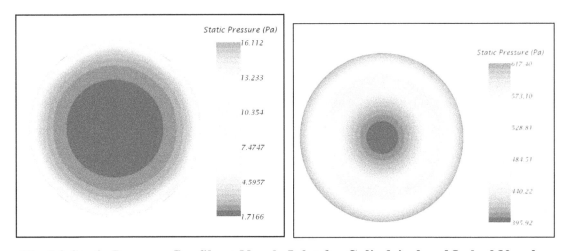

Fig 7.9 Static Pressure Profile at Nozzle Inlet for Cylindrical and Lobed Nozzles

To study the effect of nozzle shape on static pressure profile of IRSS system, maximum, minimum and average static pressures associated with case studies 2,4,5,6 and 7 are shown in Fig 7.10.

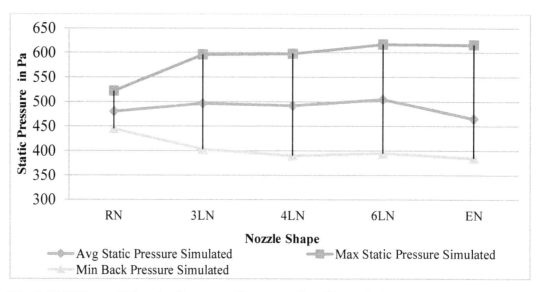

Fig 7.10 Effect of Nozzle Shape on Pressure Profile at Inlet to IRSS (Simulated)

Simulation studies of static pressure among nozzles of different shapes but same area ratio indicate that elliptical nozzle provides lower average static pressure (465 Pa) followed by round (480 Pa), 4 lobed (492 Pa), 3 lobed (496 Pa) and 6 lobed (505 Pa) nozzles. Case study 2 (RN) offers a uniform static pressure at inlet (17.3% variation) while case study 7 (EN) has maximum variation (60%). Lobed nozzles have variation ranging from 48% to 56%, increasing with the number of lobes. 4 lobed nozzle has marginally lower static pressure compared to 3 lobed and 6 lobed nozzles.

Experimental measurement of static pressure is carried out at seven different points across nozzle inlet for case study 2 (RN) and across 11 different points for case study 3 (RN1). The mass flow of exhaust plume from HAG reduced by about 50% with round nozzle and 20% with cylindrical nozzle compared to free flow without any nozzle. It is observed that the static pressure remains constant across the plane. Measured static pressure values are lower than simulated values. Simulation studies carried out with reduced mass flow indicated 80% reduction of average static pressure for 55% reduction in mass flow for round nozzle and 95% reduction in simulated average static pressure for 12% reduction in exhaust plume mass flow for cylindrical nozzle.

7.3.5 Exhaust Plume Temperature at IRSS Exit

Reduction of plume temperature is another primary design function of IRSS system. Simulation studies are carried out through case studies 2, 4, 5, and 6 for 1 kg/s exhaust mass flow at 300 °C to study the effect of nozzle shape (round, lobed or elliptical) on exhaust plume temperature at diffuser exit while Case Study 1 (RN1) illustrates the contribution of nozzle inlet to exit area ratio (A_{ni}/A_{ne}). The simulated exhaust plume temperatures at 25 equidistant points on either side of center at IRSS exit are plotted for case studies 2 to 6 in Fig 7.11.

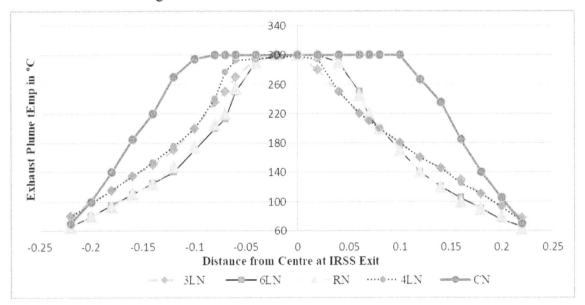

Fig 7.11 Simulated Temperature Profiles at IRSS Exit for Case Studies 2,3,4,5 & 6

Simulated exhaust plume temperature profile at IRSS exit is axisymmetric and fairly uniform at exit. The temperature increases gradually from ring wall towards the center with peak value equal to nozzle inlet temperature. It can be seen from Fig 7.11 that the slope of temperature gradient along diameter at diffuser exit is affected by the nozzle shape. Round and six lobed nozzles have similar slope while IRSS with 3 and 4 lobed nozzles have similar slope. The cylindrical nozzle has steep increase in temperatures reaching peak value closer to the wall.

Experimental studies are carried out for round and cylindrical nozzles at reduced mass flow rates due to back pressure imposed penalty. Comparison of experimental and simulated exhaust temperature profile across exit diameter is shown in Fig 7.12. The measured temperatures have a shift towards one wall (right side in graph) which is attributable to lateral shift in plume at inlet due to HAG duct profile. The measured temperatures have a lower peak indicating that ambient air is able to cool the core of the plume while simulation assumes that velocity of induced air is not sufficient to cool the core of the plume.

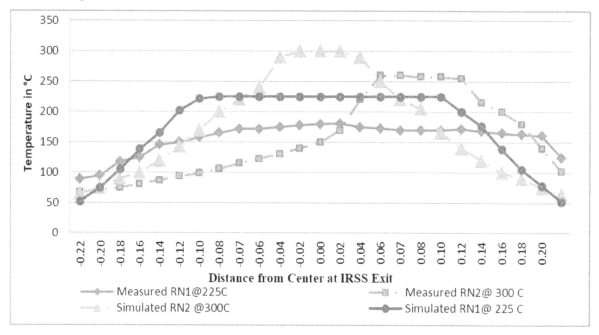

Fig 7.12 Simulated & Measured Temperature Profiles at IRSS Exit

The simulated and measured mass flow average exhaust plume temperatures for case studies 2 to 6 are indicated in Fig 7.13.

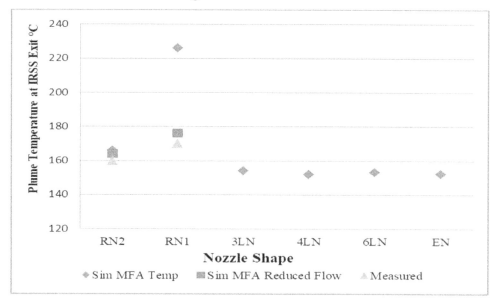

Fig 7.13 Mass Flow Average Plume Temperature at IRSS Exit – Simulated & Measured

Simulations studies indicate that convergent nozzles of case studies 2 (RN 2), 4 (3LN), 5 (4LN) and 6 (6LN) provide 46 to 50% reduction in mass flow average plume temperature whilst cylindrical nozzle (RN1) of case study 1 provides only about 25%

reduction. The enhanced cooling is attributed to increase in velocity of primary fluid to 5 m/s for cylindrical nozzle (A_{ni}/A_{ne}=1) to a peak of 50 m/s for convergent nozzles (A_{ni}/A_{ne} = 2.089) as shown in Fig 7.14. It is, however, to be noted that increase in area ratio of nozzle causes higher pressure loss and the pressure loss penalty is severe beyond area ratio of 2.13. Thus, the design of nozzle in terms of extent of reduction in exit diameter is constrained by the non-linear increase in pressure loss with area ratio. Simulation results also indicate that mass flow rate at inlet to IRSS has significant effect on reduction in plume temperature at diffuser exit for cylindrical nozzle (RN1) whilst RN2 is unaffected by the reduction in plume mass flow.

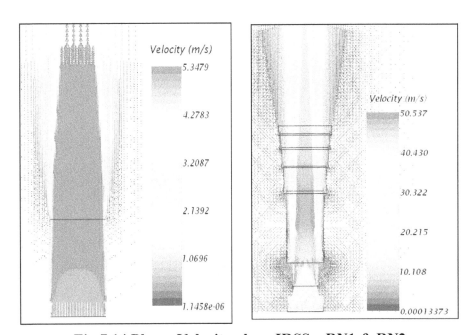

Fig 7.14 Plume Velocity along IRSS – RN1 & RN2

Experimental measurement of plume temperatures using thermocouple at 07 points across diffuser exit indicates a very high degree of correlation with mass flow average plume temperature for both convergent nozzles (Simulation- 45.33%, Measured- 46.67%) and cylindrical nozzles (Simulation-41.33%, Measured -43.33%).

7.4   MIXING TUBE DESIGN STUDY – CASE STUDY 9 & 10

IRSS devices are known in the literature to have been designed with a single mixing tube. The research work explored an innovative twin mixing tube design that maintains the same standoff and length as a single mixing tube. Performance of the twin mixing tube is compared with single mixing tube for pumping ratio, exhaust plume temperature at exit and static pressure.

7.4.1 Pumping Ratio

The secondary and tertiary pumping ratios estimated using mathematical modeling, numerical simulation and experimental measurement in respect of single and twin mixing tubes are shown in Fig 7.15 and show interesting trends.

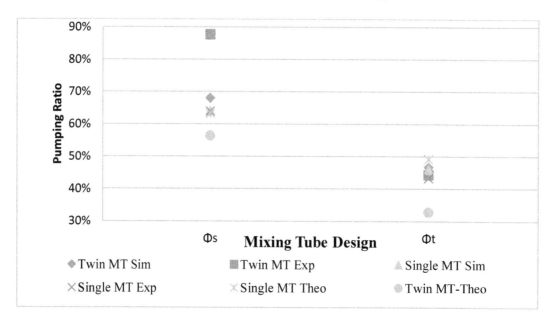

Fig 7.15 Pumping Ratio of Single & Twin Mixing Tubes - Case study 9 & 10

The mathematical model uses Eq. 4.12 given by Pritchard et al., for entrainment for a circular jet exiting from a nozzle into a mixing tube for estimating the secondary air induced into mixing tube. As seen in Fig 7.23, for the SMT design, the volume of secondary air induced in to the mixing tube estimated by the mathematical (Φs-63%) has a good match with simulation results (64%). Experimental measurements have shown 73% induction.

However, in the case of twin mixing tube, mathematical model predicted lower induction (Φs-56%) compared to simulation (68%). This is attributed to the loss of primary fluid due to expansion of the jet as it flows from bottom mixing tube to top mixing tube (of same area). Simulation results, however, have shown a positive increase (30%) in the volume of secondary air induced into the top mixing tube. Experimental

154

measurements have shown considerably higher secondary air induction for the twin mixing tube design (88%) compared to simulation (68%). Both simulation (4%) and experimental measurements (24%) have shown the twin mixing tube design to induce higher volumes of secondary air compared to single mixing tube (64%).

Tertiary air induced through the diffuser (Φ_t), estimated through simulation and experimental results for both single (43-46%) and twin (44-47%) mixing tubes, as seen in Fig 7.23, is largely unaffected by the mixing tube design. Mathematical model provides comparable estimate (Φ_t- 46%) for single mixing tube but lower estimate (33%) for twin mixing tube which is attributed to the reduced mass flow rate of primary fluid and higher temperatures associated with the flow at inlet to diffuser for twin mixing tube.

7.4.2 Mixing Tube Thermal Profile. The outer wall thermal profile of mixing tube through simulation and variation in temperature along mixing tube length obtained using a line probe are shown in Fig 7.16 for SMT and TMT designs.

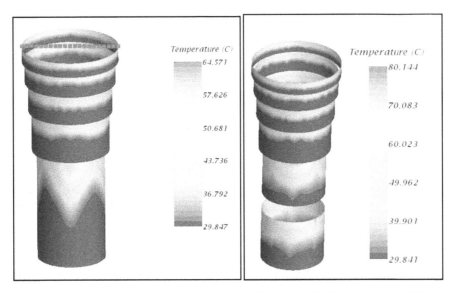

Fig 7.16 Thermal Profile of Single & Twin Mixing Tubes through Simulation

SMT has a lower thermal profile for the part corresponding to the bottom mixing tube since the thickness of air film induced into SMT for the corresponding section is thicker.

Infrared imaging of the outer wall of single and mixing tubes was carried out during experimental trials and the profiles along length of mixing tube are shown in Fig 7.17 and 7.18 respectively.

155

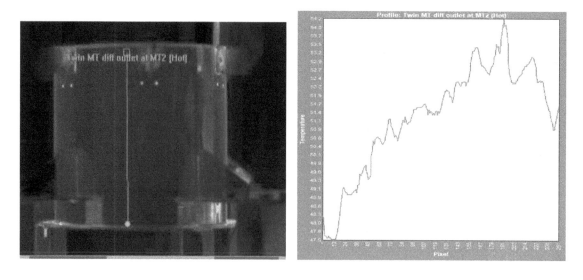

Fig 7.17 Temperature along the length of Twin Mixing Tubes (Top Half)

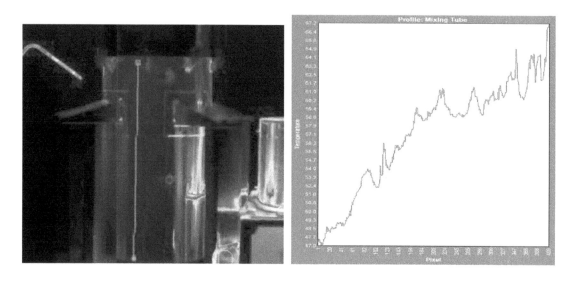

Fig 7.18 Temperature along Length of Single Mixing Tube

The experimental measurements validate the trends obtained through simulation. SMT has a peak temperature of 66 °C and surface average temperature of 56 C. The twin air induction design keeps temperature of each ring lower with a peak of 54 C and surface average temperature of 50 C. The lower temperatures associated with Twin mixing tube enable higher view angle protection and thus, provide better protection against detection by missile IR seekers.

7.4.3 Uniformity of Exhaust Plume Profile at IRSS Exit

The simulated distribution of temperature across diameter at the exit plane of IRSS for SMT and TMT are shown in Fig 7.19 (a) & (b) respectively.

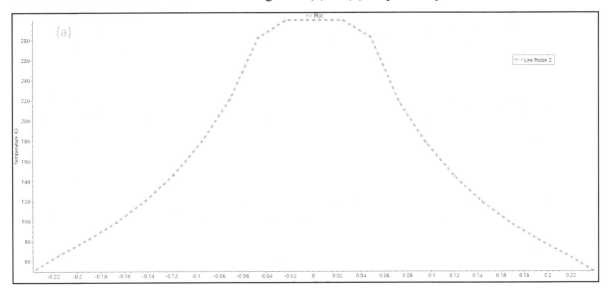

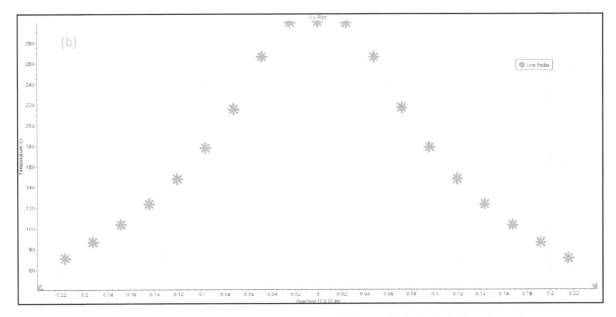

Fig 7.19 Temperature across Exit Plane for (a) Single & (b) Twin Mixing Tubes

Experimental measurements indicted that twin mixing tube design has marginally lower mass flow average plume temperature at exit (151 °C) compared to single mixing tube design (157 °C). However, SMT has a lower standard deviation (σ=5.2) for exit plume temperature compared to TMT (σ=6.5). This indicates that SMT has a comparatively uniform thermal profile at exit but with higher mean.

7.5   DIFFUSER DESIGN STUDY – CASE STUDY 2, 10,11 & 12

Diffuser design is studied to evaluate the effect of number of rings through case study 2 (4 ring diffuser) and case study 9 (2 ring diffuser). Overlap between adjacent rings is described in terms of their ring gap and is studied through case studies 9 (one ring gap overlap), 10 (No overlap) and 11 (2 ring gap overlap). 2-D drawings showing overlap between the rings and mixing tube for the four cases are shown in Fig. 7.20.

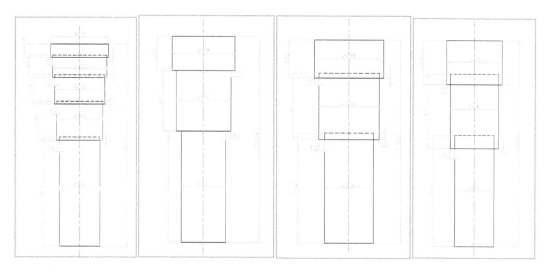

Fig. 7.20 Diffuser Design with Varying Number of Rings & Overlap between Rings

Performance parameters studied include pumping ratio, ring thermal profile and plume profile at IRSS exit. Experimental models used are shown in Fig 7.21.

7.5.1 Pumping Ratios for Different Diffuser Designs

Simulated tertiary air pumping ratios (Φt) for diffuser with varying overlap are 51%, 49% and 48% for zero, one ring gap and two ring gap overlaps respectively. Experimental measurements of tertiary air pumping ratios (Φt) for diffuser with zero, one ring gap overlap are 57% and 49% respectively. Observations that diffuser rings without overlap entrain more air are in line with literature.

Fig. 7.21 Experimental Models of Case Studies 2, 9 and 10 for Diffuser Design Study

For a 4 ring diffuser with one ring gap overlap, simulation predicted tertiary air (Φt) entrained is 46% while experimental measurement is 49%. The mass flow rate of tertiary air entrained is a function of area of ring gap and velocity of ambient air being induced. Thus, the pumping ratio of a diffuser is independent of the number of rings subject to maintaining the total area for inducing ambient air constant.

7.5.2 Ring Surface Temperatures for Different Diffuser Designs

The effectiveness of the film formed adjacent to the wall due to coanda effect within IRSS diffuser has not been discussed in the literature. Ring surface temperatures of through experimental measurement and simulation for 4 Rings Diffuser of Case Study 2 are shown in Fig 7.22. The top ring thermal profiles recorded during experimental trials of case study 9 -One Ring Gap overlap (1RG) and Case study 10 - Zero Overlap (0RG) are shown in Fig 7.23.

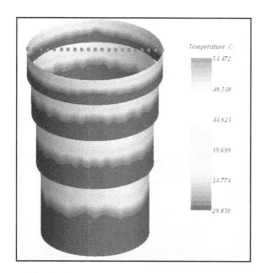
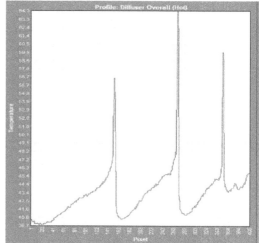

Fig. 7.22 Simulated Thermal Profile Vs Measured IR Image Plot – Case Study 2

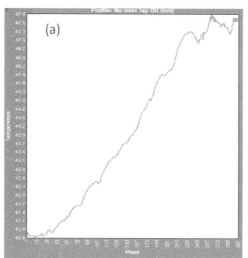
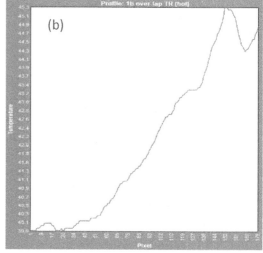

Fig. 7.23 Measured IR Profile of Diffuser Top Ring Temperatures – (a) 1RG Vs (b) 0RG

Simulated ring temperatures for case studies 2, 9, 10 and 11 are shown in Table 4.

Table 7.3. Ring Surface Average Temperatures for Diffuser Design Study

Diffuser Design	Φt	Ring 1	Ring 2	Ring 3	Ring 4
4 Rings Diffuser -1RG	47%	33.3 °C	36.7 °C	35.5 °C	35.5 °C
2 Rings Diffuser - 2RG	48%	-	38 °C	-	38.5 °C
2 Rings Diffuser - 1RG	49%	-	39 °C	-	40 °C
2 Rings Diffuser - 0RG	51%	-	40 °C	-	42.5 °C

160

It is to be noted that the total diffuser length is kept the same (1224 mm) as also the exit diameter (478 mm). The length of rings has been increased to cater for the overlap. For example, the length of top ring for the 2 ring diffusers is 201 mm for no overlap condition, 232 mm for one ring gap and 263 mm for two ring gap conditions. Experimental studies also indicate that tertiary air induced is higher for no overlap condition (Φt-59%) as also associated peak ring wall temperature (47.8 °C) compared to one ring gap (45.3 °C) diffuser. This is attributed to the 15% increase in ring length. As the tertiary air progresses along the wall, the thickness reduces and convective heat transfer from hot to cold fluid increases resulting in marginally higher wall temperatures.

Thus, it is observed that the enhanced air entrainment through no overlap condition is associated with marginally higher diffuser ring wall temperatures due to reduced effectiveness of the coanda film thickness.

7.5.3 Exhaust Plume Temperature at Diffuser Exit

Simulation studies indicated the plume temperature at diffuser exit to remain the same for all three cases of overlap (156 °C) while it is marginally higher (160 °C) for 4 rings diffuser. This is attributed to the shorter time available for mixing between induced tertiary air and primary fluid. Experimental measurements are in agreement with simulation studies.

7.6   INTEGRATED INFRARED SIGNATURE SUPPRESSION DESIGN

The performance of integrated IRSS systems vis-à-vis uniform duct associated with non-IRSS exhaust system is studied. In the case of non-IRSS exhaust system, the exit of engine is a continuous uniform diameter pipe that passes through the funnel till exit. On the other hand, passive IRSS is a structural modification at the top end of the exhaust system with an ejector and a diffuser as shown in Fig. 7.24.

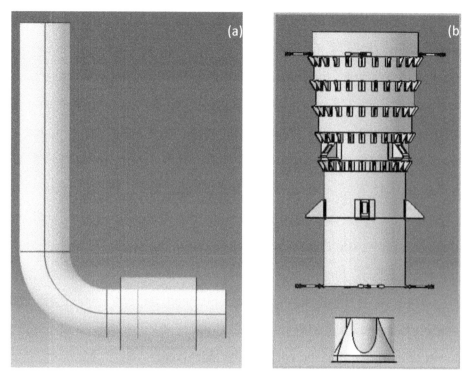

Fig. 7.24 Performance study of (a) non-IRSS (b) passive IRSS

7.6.1 Exhaust duct temperatures

Exhaust duct temperatures of diffuser rings and mixing tube are considered. In case of non-IRSS, the full length of exhaust duct is uncooled on the inner side and hence, the minor drop in temperature observed is due to flow losses. The simulated exhaust thermal profile of exhaust duct without and with IRSS along with experimental IR image of diffuser for case study 2 is shown in Fig. 7.25. The surface average temperature of non-IRSS exhaust duct is 290 °C (563 K) while that of diffuser of case study 2 is 36 .7 °C (309.7 K).

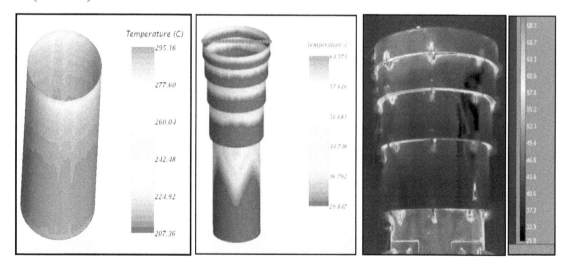

Fig. 7.25 Simulated Thermal Profile of Exhaust Without IRSS and With IRSS

7.6.2 Exhaust Plume Temperature

The simulated exhaust duct exit thermal profiles without IRSS (Case Study 12) and with IRSS (Case Study 2) are shown in Fig. 7.26. Comparison of mass flow average temperatures at exit for non-IRSS (300 °C) condition with IRSS (160 °C) indicates considerable reduction in plume temperature with IRSS. Experimental measurement of average exhaust plume temperature using 7 thermocouples across exit plane is same as simulated temperature of 160 °C.

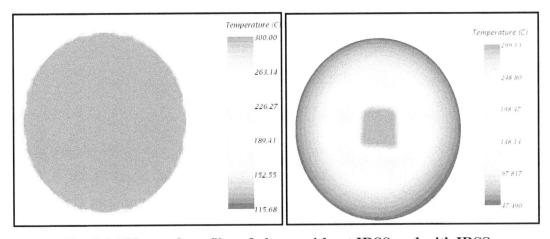

Fig. 7.26 Thermal profiles of plume without IRSS and with IRSS

For the same mass flow average plume temperature (160 °C), numerical simulation thermal profile at device exit has a high thermal spread from wall region to core (47.5 °C -299 °C) compared to experimental measurements (70 °C to 260 °C).

7.6.3 Back Pressure

Back pressure on the engine for exhaust without IRSS is negligible while IRSS system imposes pressure loss depending upon nozzle design. Mathematical model ATPM developed as part of the study helps in preliminary estimation of back pressure on the engine.

7.6.4 % Reduction in Missile Lock-on Range

Infrared signature of ship's exhaust system is predominantly the net radiance emitted by the grey body given by Stefan-Boltzmann law. The lock on range (LOR) of a missile depends on a number of factors such as detector sensitivity, directivity of detector, signal to noise ratio, diameter and numerical aperture of optics, target radiance, background radiance (Schleher,1999; Mohammed Syuhaimi et al., 2009).

For a given a set of missile seeker parameters and target area, lock on range is dependent only on target radiance (N_t) (target t_1 is exhaust with IRSS and target t_0 is exhaust without IRSS) and background radiance (N_b). % change in LOR is given by

$$\% \text{ change in LOR} = \frac{\sqrt{(N_{t1}-N_b)} - \sqrt{(N_{t0}-N_b)}}{\sqrt{(N_{t0}-N_b)}} \times 100 \qquad (8.1)$$

The emissivity of sea background is considered as $\varepsilon = 0.984$ (Konda et al., 1994). For the temperatures associated with case Studies 2 and 12, the % reduction in missile LOR for exhaust system with and without IRSS is **80.31%**.

7.7 CONCLUSION

The chapter groups the case studies based on design of key subsystem and analyses the effect of component design on the performance of integrated IRSS system in terms of pumping ratios, fluid parameters at device exit, diffuser ring thermal profile, back pressure due to IRSS and Missile Lock-On Range with respect to the baseline infrared signature associated with non-stealth ships. It is concluded that from the study on infrared signature suppression for shipboard applications that Stealth (with IRSS) ships fitted with hypothetical exhaust systems as considered in Case Study 2 have lower IR signature compared to Non-Stealth ships without IRSS (Case Study 12). The % reduction in Missile Lock-On range is up to 80.31%.

CHAPTER 8

CONCLUSIONS AND FUTURE SCOPE

8.1   INTRODUCTION

Salient contributions of the research work and conclusions drawn from the research work are presented. Future scope of work both in terms of enhancing available literature on passive IRSS technology as also in exploring alternate technologies for holistic management of full ship infrared signature is recommended.

8.2   CONCLUSIONS

The research work studied the available literature and observed technology gaps in predicting the exhaust plume flow characteristics when a combination of ejectors and diffusers are installed in the exhaust systems onboard ships. To facilitate the design of an integrated ejector-diffuser infrared suppressor for ship specific funnel space constraints, mathematical model 'Ejector-Diffuser for Ship Applications (EDSA)' has been developed along 'Aero-Thermal Performance (ATPM) model for predicting the flow properties and back pressure on the engine for generated dimensional designs.

Numerical simulation studies are carried out to identify the effect of component level design variations (03 nozzle exit areas, 05 nozzle shapes, 02 mixing tube designs, 02 diffuser ring designs, three overlap conditions) on the performance of an Integrated Ejector-Diffuser with respect to infrared suppression. Select models are fabricated and experimental trials are undertaken using a hot air generator at naval Science and Technology Laboratory, Visakhapatnam to validate the results obtained through numerical simulation and ATPM.   Salient contributions of the research work and conclusions drawn are as follows:-

(i)   Study of secondary air entrainment due to nozzle exit area (all round shape) and nozzle shapes (A_{ne}/A_{ni} of 2.1 for all) indicated that Nozzle RN 3 (A_{ne}/A_{ni} of 3.1) entrained maximum (173%) ambient air while nozzle RN1 (A_{ne}/A_{ni} of 1.1) entrained minimum (28%) ambient air. Entrainment due to different nozzle shapes varied from 110 % (RN2) to 121% (4LN).

(ii)   Nozzle exit shape (round/lobed/elliptical) has increased the pumping ratio through standoff for various shapes (RN- 64%, 3LN - 70%; 4LN, 6LN and EN - 73%) but had reduced effect on the pumping ratio of tertiary air (Φ_t) induced through diffuser rings (4LN - 48%; RN, 3LN and 6LN - 46%).

(iii)   Experimentally measured pumping ratio of secondary air (Φ_s) for round nozzle RN 2 is higher (70%) than simulation (64%). On the other hand, experimentally measured tertiary pumping ratio (13%) is less than simulation (46%). The pumping ratio of total entrained air (Φ_T) for measured (199%) and simulated (201%) is in good agreement since higher mass flow through standoff compensates the reduced entrainment through diffuser. The same trend is observed for experimental and simulated values of RN1 also.

(iv)   Average static pressure for convergent nozzles with large area ratio (A_{ni}/A_{ne} = 2.089) is over 300 times higher than cylindrical nozzle (A_{ni}/A_{ne} = 1)

(v)   Elliptical nozzle provides lower average static pressure (465 Pa) followed by round (480 Pa), 4 lobed (492 Pa), 3 lobed (496 Pa) and 6 lobed (505 Pa) nozzles.

(vi)   Convergent nozzles of case studies 2 (RN 2), 4 (3LN), 5 (4LN) and 6 (6LN) provide 46 to 50% reduction in mass flow average plume temperature whilst cylindrical nozzle (RN1) provides about 25% reduction.

(vii)   Measurement of plume temperatures at diffuser exit showed good correlation with simulation results for both convergent (Simulation- 45.33%, Measured-46.67%) and cylindrical nozzles (Simulation-41.33%, Measured -43.33%).

(viii)   Experimental measurements have shown higher secondary air induction for the twin mixing tube design (88%) compared to simulation (68%).

(ix)   Twin Mixing Tube (TMT) has a peak metal temperature of 54 °C compared to peak of 66 °C for SMT. Surface average temperature of TMT is lower at 50 °C compared to 56 °C for SMT.

(x)   Experimental measurements indicted that TMT design has lower mass flow average plume temperatures at diffuser exit (151 °C) compared to SMT design (157 °C).

(xi)   Simulated tertiary air pumping ratios are 51%, 49% and 48% and 47% for 0RG, 1RG, and 2RG respectively. Experimental measurements recorded tertiary air pumping ratios of 57% and 49% for 0RG and 1RG respectively.

(xii)   However, integrated ejector-diffuser with 'No Overlap' emitted 9% higher infrared signature than 4 Ring Diffuser and 5% higher than 2RG. This is attributed to the 15% increase in ring length over 2GP. This highlights that optimal air entrainment does not give optimal IR suppression.

(xiii)   Total air entrained into diffuser is not significantly affected (<5%) by the number of rings (4 Ring and 2 Ring) subject to area ratio between diffuser exit and MT exit being same.

(xiv)   The surface average temperature of non-IRSS exhaust duct is 290 °C (563 K) while that of diffuser of case study 2 is 36 .7 °C (309.7 K). This corresponds to 90.84% reduction in radiant intensity due to exhaust duct wall temperatures with IR suppressor installed in ship exhaust systems

(xv)   The mass flow average exhaust plume temperature at funnel exit for non-IRSS exhaust duct is 300 °C (573 K) while that of IRSS (Case Study 2) is 160 °C (433 K). This corresponds to 67.39% reduction in radiant intensity due to exhaust plume with IR suppressor installed in ship exhaust systems

(xvi)   For the same mass flow average plume temperature (160 °C), numerical simulation thermal profile at device exit has a high thermal spread from wall region to core (47.5 °C -299 °C) compared to experimental measurements (70 °C to 260 °C).

(xvii)   For the temperatures associated with case Studies 2 and 12, the % reduction in missile LOR for exhaust system with and without IRSS is **80.31%.**

8.3 FUTURE SCOPE OF WORK

The future scope of work is recommended both in terms of enhancing available literature on passive IRSS technology as also in exploring alternate technologies for holistic infrared signature management of the complete marine platform. Specific areas recommended for further research are as follows:-

(a) Experimental validation with higher mass flow exhaust systems such as main propulsion cruise and boost systems to study effects of scaling

(b) Expand the ATPM mathematical model to take component geometric length into consideration and to predict surface average exhaust duct wall temperature

(c) Studies on optimization of alternate IRSS dimensional designs for maximizing IR signature suppression and minimizing back pressure penalty

(d) Study the effect of noise generated by the ejector-diffuser passive IRSS system especially for high exhaust mass flow systems such as plume turbines;

(e) Carryout simulation and experimental studies with water mist of varying droplet diameters for cooling exhaust plume

8.4 CONCLUSION

Salient contributions of the research work and conclusions drawn from the research work are presented. Future scope of work both in terms of enhancing available literature on passive IRSS technology as also in exploring alternate technologies for holistic management of full ship infrared signature is recommended.

REFERENCES

Ab-Rahman, M. S., & Hassan, M. R. (2009). Lock-on range of infrared heat seeker missile. International Conference on Electrical Engineering and Informatics, Vol. 2, pp: 472-477.

Ashok, K.B., & Sukanta K. D., Pandaba P., Subrat M. (2014). Experimental and numerical investigation of air entrainment into a louvred funnel. Applied Ocean Research, Vol. 48, pp:176-185.

Baham, G. J., & McCallum, D. (1977). Stack design technology for naval and merchant ships. SNAME Transactions, Vol. 85, pp:324-349.

Barik, A. K., Dash, S. K., & Guha, A. (2015). Experimental and numerical investigation of air entrainment into an infrared suppression device. Applied Thermal Engineering, Vol. 75, pp:33-44.

Bhattacharya, A., & Achari, A. M. (2020). Fluid flow analysis through a variable diameter infrared suppression device. International Journal of Automotive and Mechanical Engineering, Vol. 17(1), pp:7734-7748.

Birk, A. M., & Davis, W. R. (1989). Suppressing the infrared signatures of marine plume turbines. J Eng Gas Turbines Power, Vol. 111(1), pp:123-129.

Birk, A. M., & VanDam, D. (1989). Marine gas turbine infra-red signature suppression: Aerothermal design considerations. Proc.ASME.GT1989, Vol 2, pp: 002-009.

Birk, A. M., and VanDam, D.(1994). Infrared Signature Suppression for Marine Gas Turbines: Comparison of Sea Trial and Model Test Results for the DRES Ball IRSS System. ASME. *J. Eng. Gas Turbines Power*; Vo. 116(1): pp:75–81.

Bonnington, S. T., King, A. L., & Hemmings, J. A. G. (1972). Jet Pumps and Ejectors: A state of the art review and bibliography.

Carletti, M. J., Rogers, C. B., & Parekh, D. E. (1995). Use of streamwise vorticity to increase mass entrainment in a cylindrical ejector. AIAA journal, Vol. 33(9), pp:1641-1645.

Center, Naval Air Warfare. (1997). Electronic warfare and radar systems engineering handbook. Electronic Warfare Division, Pont Mugu, CA.

Chandrakar, V., Mukherjee, A., Senapati, J. R., & Mohanty, A. (2022). Conjugate free convection with surface radiation from real-scale IRS system with multiple conical funnels: International Journal of Thermal Sciences, Vol.153, pp:106004

Chen, Q., & Birk, A. M. (2008). Experimental Study of Oblong Exhaust Ejectors With Multi-Ring Oblong Entraining Diffusers. In Turbo Expo: Power for Land, Sea, and Air, Vol. 43161, pp:11-21.

Chen, Z., & Yao, G. (2019). Analysis of Radar and Infrared Stealth Compatibility Design for Surface Ships. IOP Conference Series: Earth and Environmental Science. Vol. 252 (5), p. 052132.

Cho, Y. J., & Ko, D. E. (2017). A Study on the Characteristics of Design Variables for IRSS Diffuser. IOP Conference Series: Materials Science and Engineering. Vol. 269 (1), pp:012092.

Choudhary, K. K., Kumar, P. G., & Murty, A. V. S. N. (2022). Infrared Suppression Technologies for Marine Platforms. Journal of Institution of Engineers (India): Series C, **Volume 103, Issue 3, pp:413-419**

Collins, J. T. (1975). USS Spruance (DD963) Class, Designed for Change—The Payoff is Now. Naval Engineers Journal, Vol.87(2), pp:27-31.

Darling RS, 1973, "Feasibility Study of Eductors Applied to Naval Ship Stack," Boeing Company, Vertol Division Report Number D210-10650-l.

Dash, M. K., & Barik, A. K. (2023). A computational study on air-entrainment and pressure distribution for natural convection cooling of a hybrid IRS device. International Journal of Thermal Sciences, Vol.187, pp:108196.

Davis, W. R., & Thompson, J. (2002). Developing an IR signature specification for military platforms using modern simulation techniques. Ottawa, Ontario, Canadá. [On line: http://www. davis-eng. com/docs/papers/IR% 20Signature% 20Specification. Pdf].

Decher, R. (1981). Infrared emissions from turbofans with high aspect ratio nozzles. Journal of Aircraft, Vol.18(12), pp:1025-1031.

Dipti, P.M & Subhash,C.P. (2019). Computation of Air Entrainment into a Mixing Pipe: An Experimental and Numerical Analysis. Journal of Mechanical Engineering Vol 16(2).

Dudar, S. W. (2003). Preliminary design study of an enhanced mixing eductor system for the LHA (R) plume turbine exhaust. Naval Postgraduate School, Monterey CA.

Dvořák, V. (2007). Study of optimization of lobed nozzle for mixing. Colloquium Fluid Dynamics, Institute of Thermo mechanics AC CR, Prague, Czech Republic.
Ellin, C. R., & Pucci, P. F. (1977). Model Tests of Multiple Nozzle Exhaust Plume Eductor Systems for Plume Turbine Powered Ships. Naval Postgraduate School Monterey CA.

Faqihi, B., & Ghaith, F. (2023). A comprehensive review and evaluation of heat recovery methods from plume turbine exhaust systems. International Journal of Thermofluids, Vol.18, pp:100347.

Fitzgerald, M. P. (1986). A method to predict stack performance. Naval engineers journal, Vol. 98(5), pp35-46.

Galle, L. F., & Schleijpen, H. M. A. (1998). Ship Survivability (part II) Ship Infrared (IR) signatures.

Ganguly Rao, Arvind. (2011).Infrared signature modelling and analysis of aircraft plume. International Journal of Turbo Jet Engines.28.187-197.10.1515/tjj.2011.023.Aa

Ganguly, V. R., & Dash, S. K. (2020). Numerical analysis of air entrainment and exit temperature of a real scale conical infrared suppression (IRS) device. International Journal of Thermal Sciences, Vol.156, pp:106482.

GER-4250 - GE's LM2500+G4 Aeroderivative Plume Turbine for Marine and Industrial Applications, accessed on 01 June 2023

Hamedi-Estakhrsar, M. H., Mahdavy-Moghaddam, H., & Jahromi, M. (2018). Investigation of effects of convergence and divergence half-angles on the performance of a nozzle for different operating conditions. Journal of the Brazilian Society of Mechanical Sciences and Engineering, Vol.40, pp1-12.

Hiranandani, V. A. G. (2009). Transition to Guardianship: The Indian Navy 1991–2000. Lancer Publishers LLC.

Ho, C. M., & Gutmark, E. (1987). Vortex induction and mass entrainment in a small-aspect-ratio elliptic jet. Journal of Fluid mechanics, Vol.179, pp:383-405.

Hoffmann, J. A., & Gonzalez, G. (1984). Effects of small-scale, high intensity inlet turbulence on flow in a two-dimensional diffuser. ASME, Transactions, Journal of Fluids Engineering, 106.

https://indiandefenseanalysis.com/2023/01/28/indian-navy-has-facilitated-a-mou-towards-indigenous-manufacture-of-mtu-4000-series-marine-diesel-engines; accessed 01 June 2023

https://www.mod.gov.in/dod/technology-perspective-and-capability-roadmap-tpcr-2018; accessed on 01 June 23

https://www.mordorintelligence.com/industry-reports/military-electro-optical-and-infrared-systems-market; accessed 01 June 23 https://www.mtu-solutions.com/eu/en/applications/defense/marine-defense-solutions.html; accessed 01 June 23

Hudson, R. D. (1969). Infrared system engineering. Wiley-Inter science. Vol. 1, pp: 642.

Hyun Im, J., & Jin Song, S. (2015). Mixing and entrainment characteristics in circular short ejectors. Journal of Fluids Engineering, Vol.137(5).

Johnston, I. H. (1953). The effect of inlet conditions on the flow in annular diffusers. National Plume Turbine Establishment Farnborough (United Kingdom).

Klein, A. (1981). Effects of inlet conditions on conical-diffuser performance.

Kok, S. L. (2012). Naval survivability and susceptibility reduction study-surface ship. Naval Postgraduate School, Monterey, CA.

Kulkarni, P. R., Singh, S. N., & Seshadri, V. (2005). Study of smoke nuisance problem on ships–a review. International Journal of Marine Engineers, Proceedings of Royal Institute Naval Architecture Part A2, Vol.147, pp.27-50.

Kutzscher, Edgar, 1957, "The physical and technical development of infrared homing devices", Technical report No. AGARD-AG-20, History of German Guided Missiles Development, NATO .

Liu, Y. H. (2002). Experimental and numerical investigation of circularly lobed nozzle with/without central plug. International Journal of Heat and Mass Transfer, Vol.45(12), pp.2577-2585.

LM2500 & LM2500XPRESS Plume Turbines | GE Plume Power, accessed 08 May 23.

Maheshwaran, M., Praveen K.D. & Vinothkumar, V.S. (2020). Numerical Investigation of flow through conical diffuser with Swirl Generator. Turkish Journal of Computer and Mathematics Education, Vol.11, No.03.

Manoj, K.D., Sukanta,K.D. (2019). 3D numerical study of natural convection heat transfer from a hollow horizontal cylinder placed on the ground. International Journal of Thermal Sciences Vol.140,pp.429–441.

Maqsood, A., & Birk, A. M. (2007, January). Effect of a Bend on the Performance of an Oblong Ejector. In Turbo Expo: Power for Land, Sea, and Air.Vol. 47950, pp. 37-45.

McDonald, A. T., Fox, R. W., & Van Dewoestine, R. V. (1971). Effects of swirling inlet flow on pressure recovery in conical diffusers. AIAA Journal, Vol.9(10), pp.2014-2018.
Mi, J., Nathan, G. J., & Luxton, R. E. (2000). Centreline mixing characteristics of jets from nine differently shaped nozzles. Experiments in Fluids, Vol.28(1), pp.93-94.

Mishra, D. P., & Dash, S. K. (2010). Prediction of entrance length and mass suction rate for a cylindrical sucking funnel. International journal for numerical methods in fluids, 63(6), pp.681-700.

Mishra, S. K., Barik, A. K., & Swain, P. K. (2022). Air Entrainment and Outlet Temperature Characteristics of a Modified Infrared Suppression Device With Inward and Outward Guides. Journal of Thermal Science and Engineering Applications, Vol.14(12), 121006.

Mitchell, J. W. (1958). Design parameters for subsonic air-air ejectors. Stanford University. Dept Of Mechanical Engineering.

Mohammadaliha, N., Afshin, H., & Farhanieh, B. (2016). Numerical investigation of nozzle geometry effect on turbulent 3-D water offset jet flows. Journal of Applied Fluid Mechanics, Vol.9(4), pp.2083-2095.

Mueller, N. H. G. (1964). Water jet pump. Journal of the Hydraulics Division, Vol.90(3), pp.83-113.

Mukherjee D.K. (1976). Film Cooling with Injection through Slots. Transactions of ASME. Pp.556-559

Munro, G., VanDam, D., & Birk, A. M. (1990). The Design of Infrared Signature Suppression Hardware for the Refit of Canada's DDH 280 Destroyers. Transactions of ASNE.

Neele, F. P., Wilson, M. T., & Youern, K. (2005). Proposed New Method of Interpretation of Infrared Ship Signature Requirements. Naval Forces, Vol.26(6),pp. 32.

Nichols, L.W., et al., 1959 "Military Applications of Infrared Techniques," in Proceedings of the IRE, vol. 47, no. 9, pp. 1611-1624,

Nicoll, W. B., & Ramaprian, B. R. (1970). Performance of conical diffusers with annular injection at inlet.

Prichard, R., Guy, J. J., & Conner, N. E. (1977). Industrial plume utilization. New Providence (NJ): Bowker.

Pucci, P. F. (1955). Simple Ejector Design Parameters. Stanford University.

Rao, A. G. (2011). Infrared signature modeling and analysis of aircraft plume.

Reddy, Y. R., & Kar, S. (1968). Theory and performance of water jet pump. Journal of the Hydraulics Division, Vol.94(5), pp.1261-1282.

Reddy, Y. R., & Kar, S. (1968). Theory and performance of water jet pump. Journal of the Hydraulics Division, Vol.94(5), pp.1261-1282.

Rodrigo, Q.N. & Aldo J.D., & Jose G.C., & Antonio, C.P., & Brasil, J. (2007). Experimental and Numerical Study of the Swirling Flow in Conical Diffusers. 19th International Congress of Mechanical Engineering, Brasilia.

Sahu, S. R., & Mishra, D. P. (2015). Effect of Pipe Configurations on Air Entrainment into a Louvered Cylindrical Pipe: A Comparison between Open and Close Entrance of a Pipe. International Journal of Engineering Research, Vol.4(4), pp.173-177.

Schleher, D. C. (1999). Electronic warfare in the information age. Artech House, Inc.

Schleijpen, H. M. A., & Neele, F. P. (2004, August). Ship exhaust plume plume cooling. In Targets and Backgrounds X: Characterization and Representation. Vol. 5431, pp. 66-76.

Sen, S. (2008). Studies on the flow characteristics of a stepped conical diffuser with passive suction (Doctoral dissertation).

Senoo, Y., Kawaguchi, N., & Nagata, T. (1978). Swirl flow in conical diffusers. Bulletin of JSME, Vol.21(151), pp.112-119.

Shan, Y., & Zhang, J. Z. (2009). Numerical investigation of flow mixture enhancement and infrared radiation shield by lobed forced mixer. Applied Thermal Engineering, Vol.29(17-18), pp.3687-3695.

Singh L., & Singh S.N., & Sinha S.S, (2017). Flow and Heat transfer characteristics of a self-entraining Ejector-Conical Diffuser with and without slot guidance. 13th International Conference on Heat Transfer, Fluid Mechanics and Thermodynamics.

Singh, G., Sundararajan, T., & Bhaskaran, K. A. (2003). Mixing and entrainment characteristics of circular and noncircular confined jets. J. Fluids Eng., Vol.125(5), pp.835-842.

SINGH, H., & ARORA, B. (2019). Effects of inlet conditions on diffuser performance. Int. J. Mech. Prod. Eng. Res. Dev., Vol.9, pp.813-832.

Singh, L., Singh, S. N., & Sinha, S. S. (2018). Effect of standoff distance and area ratio on the performance of circular exhaust ejector using computational fluid dynamics. Proceedings of the Institution of Mechanical Engineers, Part G: Journal of Aerospace Engineering, Vol.232(15), pp.2821-2832.

Singh, L., Singh, S. N., & Sinha, S. S. (2019). Effect of slot-guidance and slot-area on air entrainment in a conical ejector diffuser for infrared suppression. Journal of Applied Fluid Mechanics, Vol.12(4), pp.1303-1318.

Singh, L., Singh, S. N., & Sinha, S. S. (2019). Effect of slot-guidance and slot-area on air entrainment in a conical ejector diffuser for infrared suppression. Journal of Applied Fluid Mechanics, Vol.12(4), pp.1303-1318.

Singh, L., Singh, S. N., & Sinha, S. S. (2021). Enhancement of air entrainment in ejector-diffuser using plate guidance at slots to reduce infrared emission. Proceedings of the Institution of Mechanical Engineers, Part G: Journal of Aerospace Engineering, Vol.235(10), pp.1284-1305.

Singh, P., & Singh, S. N. (2014). Effect of number of slots and overlap on the performance of non-circular ejector air diffuser. In 32nd AIAA applied aerodynamics conference p. 2837.

Singh, P., & Singh, S. N. (2014). Effect of number of slots and overlap on the performance of non-circular ejector air diffuser. In 32nd AIAA applied aerodynamics conference.p. 2837.

Singh, P., Singh, S., & Seshadri, V. (2009, June). Experimental Investigations on non-circular ejector air diffusers. In 39th AIAA fluid dynamics conference.p. 4213.

Singh, S. N., Agrawal, D. P., Sapre, R. N., & Malhotra, R. C. (1994). Effect of inlet swirl on the performance of wide-angled annular diffusers.

Singhal, T., Singh, S. N., Mathur, S., & Singh, R. K. (2006). Performance optimization for two-dimensional rectangular diffuser by momentum injection using computational fluid dynamics. Proceedings of the Institution of Mechanical Engineers, Part C: Journal of Mechanical Engineering Science, Vol.220(12), pp.1775-1783.

Skebe, S., Mc Cormick, D., & Presz, JR, W. (1988). Parameter effects on mixer-ejector pumping performance. In 26th Aerospace Sciences Meeting (p. 188).Hui Hu, et al. Research on the rectangular lobed exhaust ejector/mixer systems. Trans-Jpn Soc Aeronaut Space Sci 1999; Vol.41: pp.187–194.

Sparrow, E. M., Abraham, J. P., & Minkowycz, W. J. (2009). Flow separation in a diverging conical duct: Effect of Reynolds number and divergence angle. International Journal of Heat and Mass Transfer, Vol.52(13-14), pp.3079-3083.

Thompson, J., Vaitekunas, D., & Birk, A. M. (1998, January). IR signature suppression of modern naval ships. In ASNE 21st century combatant technology symposium. Vol. 1, pp. 27-30.

Tsan-Hsing, S., William, W.L., Aamir S., ZhigangYang., & Jiang Zhu., (1995). A New k-ε Eddy Viscosity Model for high Reynolds Number Turbulent Flows. Computers Fluids Vol.24, No.3,pp. 227-238.

Vijayakumar, R., Singh, S. N., Seshadri, V., & Kulkarni, P. R. (2012). Flow visualization studies of exhaust smoke interaction with superstructure and intakes of generic naval ships in a wind tunnel. Naval Engineers Journal, Vol.124(3), pp.129-142.

Vyas, B. D., & Kar, S. (1975). Study of entrainment and mixing process for an air to air jet ejector. In In: Symposium on Jet Pumps and Ejectors and Plume Lift Techniques. pp. C2:15-C2:25.

Waitman, B. A., Reneau, L. R., & Kline, S. J. (1961). Effects of inlet conditions on performance of two-dimensional subsonic diffusers. Vol.83, pp:349-360.

Wang, S. F., & Li, L. G. (2006). Investigations of flows in a new infrared suppressor. Applied thermal engineering, Vol.26(1), pp.36-45.

Wilsted, H. D., Huddleston, S. C., & Ellis, C. W. (1949). Effect of temperature on performance of several ejector configurations. Report No. NACA-RM-E9E16.

Wright, J. R. (2020). Static Pressure Recovery Effects of Conical Diffusers with Swirling Inlet Flow. Doctoral dissertation, Clemson University. Static Pressure Recovery Effects of Conical Diffusers with Swirling Inlet Flow (core.ac.uk) accessed 01 June 23.

Yong-Jin C., & Dae-Eun K., (2017). A Study on the Characteristics of Design Variables for IRSS Diffuser. IOP Conf. Series: Materials Science and Engineering Vol.269. p.012092

Milton Keynes UK
Ingram Content Group UK Ltd.
UKHW032228120324
439302UK00013B/740